## Studies in Systems, Decision and Control

Volume 284

**Series Editor**

Janusz Kacprzyk, Systems Research Institute, Polish Academy of Sciences, Warsaw, Poland

AF904186

The series "Studies in Systems, Decision and Control" (SSDC) covers both new developments and advances, as well as the state of the art, in the various areas of broadly perceived systems, decision making and control–quickly, up to date and with a high quality. The intent is to cover the theory, applications, and perspectives on the state of the art and future developments relevant to systems, decision making, control, complex processes and related areas, as embedded in the fields of engineering, computer science, physics, economics, social and life sciences, as well as the paradigms and methodologies behind them. The series contains monographs, textbooks, lecture notes and edited volumes in systems, decision making and control spanning the areas of Cyber-Physical Systems, Autonomous Systems, Sensor Networks, Control Systems, Energy Systems, Automotive Systems, Biological Systems, Vehicular Networking and Connected Vehicles, Aerospace Systems, Automation, Manufacturing, Smart Grids, Nonlinear Systems, Power Systems, Robotics, Social Systems, Economic Systems and other. Of particular value to both the contributors and the readership are the short publication timeframe and the world-wide distribution and exposure which enable both a wide and rapid dissemination of research output.

** Indexing: The books of this series are submitted to ISI, SCOPUS, DBLP, Ulrichs, MathSciNet, Current Mathematical Publications, Mathematical Reviews, Zentralblatt Math: MetaPress and Springerlink.

More information about this series at http://www.springer.com/series/13304

Isabella Corradini

# Building a Cybersecurity Culture in Organizations

## How to Bridge the Gap Between People and Digital Technology

Springer

Isabella Corradini
Themis Research Center
Rome, Italy

ISSN 2198-4182 ISSN 2198-4190 (electronic)
Studies in Systems, Decision and Control
ISBN 978-3-030-43998-9 ISBN 978-3-030-43999-6 (eBook)
https://doi.org/10.1007/978-3-030-43999-6

© Springer Nature Switzerland AG 2020
This work is subject to copyright. All rights are reserved by the Publisher, whether the whole or part of the material is concerned, specifically the rights of translation, reprinting, reuse of illustrations, recitation, broadcasting, reproduction on microfilms or in any other physical way, and transmission or information storage and retrieval, electronic adaptation, computer software, or by similar or dissimilar methodology now known or hereafter developed.
The use of general descriptive names, registered names, trademarks, service marks, etc. in this publication does not imply, even in the absence of a specific statement, that such names are exempt from the relevant protective laws and regulations and therefore free for general use.
The publisher, the authors and the editors are safe to assume that the advice and information in this book are believed to be true and accurate at the date of publication. Neither the publisher nor the authors or the editors give a warranty, expressed or implied, with respect to the material contained herein or for any errors or omissions that may have been made. The publisher remains neutral with regard to jurisdictional claims in published maps and institutional affiliations.

This Springer imprint is published by the registered company Springer Nature Switzerland AG
The registered company address is: Gewerbestrasse 11, 6330 Cham, Switzerland

# Introduction

Over the last few years, cybersecurity has become one of the most important issues for organizations across all sectors, industries and countries. In fact, an effective management of cybersecurity ensures every organization the protection of its business and reputation.

Cyberthreats and actors are becoming more and more sophisticated, especially because of the advancement of technology, such as Internet of Things (IoT) and Artificial Intelligence (AI). Institutions and companies are improving their capabilities to strengthen cyber resilience, but results are far from being effective. To achieve more positive outcomes, they need to revise their approach, currently mainly based on technological solutions, and to invest in prepared and responsible staff.

Cybersecurity is typically associated with experts in Information Technology (IT), because the concept calls on technical aspects. Actually, cybersecurity is about the relation between people and digital technologies, even though individuals often delegate cybersecurity issues, convinced that this subject does not concern them.

Nevertheless, cybersecurity cannot be delegated, because of the social and behavioural implications: According to the most important cybersecurity reports, people are an essential part of the security problem.

Many examples show how unwisely people use digital technology: most of them still use weak passwords, easy to remember but also simple to crack; they love sharing data and information on social media; they continue to open emails from unknown recipients trusting in their ability to avoid negative consequences. We have to admit that phishing is still a popular threat, and it works precisely because it is strictly related to human nature: Exploiting people's weaknesses, like curiosity and trust, and using a persuasive communication, phishing emails continue to be successful.

Individuals lack adequate sensors to confidently act in the digital world, and they are not aware of the risks they are exposed to. This is why cybersecurity education is a priority and must be addressed to everyone, with the goal of developing secure behaviour. Indeed, we all use digital technologies, we are immersed in them and we will be even more in the future: the responsible use of digital technologies and the protection of our information should become something natural and physiological in our daily routine.

On the other hand, the impressive evolution of technology is going to have a deep impact on our lives. Artificial Intelligence and Internet of Things will probably bring benefits for many sectors and activities, but they will be accompanied by further security and privacy risks that will have to be handled. We appreciate the advantages deriving from the use of wearable devices, but we must not forget that they are becoming particularly invasive, putting our privacy and security at risk.

Since digital evolution brings benefits for everyone, also criminals take advantage of the opportunities offered by technology, by exploiting the combination between physical and digital elements to commit a crime. In fact, even if digital crime develops in a digital environment, physical elements are often involved, as it happens, for example, with the Internet of Things. Here, the multiplication of data represents a further opportunity for cybercriminals. At the same time, traditional crimes exploit cyber methods to strengthen their modus operandi. So, physical and cyber enhance each other.

Considering the complexity of this scenario, a holistic approach for managing security is fundamental. There is a wide consensus, in theory, on considering people, technology and process as an intertwined set of elements needed for effective cybersecurity; however, in practice, this approach is far from being implemented.

In this book, the focus is on the importance of people and on the need for bridging the gap with digital technologies, since ultimately technology is always a mediator of social interactions. Therefore, managing cybersecurity means, above all, to develop a people-centric security strategy: There can be no security without a secure behaviour, and there is no technical measure, even the most sophisticated, which can ensure a full security.

Organizations have to improve their cybersecurity posture through the building of a strong Cybersecurity Culture. They are more and more dependent on digital and advanced technologies and they need to reinforce security, because the more they are digitally interconnected the more the risks grow.

We need to redefine the approach to cybersecurity, starting from the statement that people are an essential part of the strategy and that Cybersecurity Culture is an ongoing process; to be effective, it takes time and effort. The issue, in fact, is not how to technically handle threats, but how to change people's mindset in order to make them fully aware of cyber-risks.

The best means to achieve this goal is education. When people are aware and use technologies responsibly, they can manage risks and appreciate their advantages, too. At the same time, a wider security awareness would be important to all citizens, because the real risk is a society made up of individuals who are proud to own latest generation products and to be able to turn on a thermostat remotely, but are not aware of the security, privacy and social implications of these actions. Ignoring the importance of having informed people and focusing only on IT technicians are equivalent to prepare a war by training Generals, and not Troops.

Talking about human factors, we need to consider the different dimensions: Social, cognitive, behavioural and emotional. In this book, we analyse the role of individuals related to security and how their weaknesses can be exploited by cybercriminals. However, our goal is to strengthen the role of people, overcoming the stereotypical view of human beings as the weakest link of security: When properly trained and actively involved, they can become the real line of defence of every organization, the so-called human firewall.

What about human factors in cybersecurity? Can communication be both a threat and a strategy? What about the link between cyberthreats and company's reputation? What should be done to achieve an effective Cybersecurity Culture? What methods and tools have to be used for ensuring positive outcomes?

These issues are discussed on the basis of the author's on-the-field work and academic experience, both in safety and in security areas. This book proposes a psychosocial vision on cybersecurity, based on the current literature, providing some recommendations to handle the process of building a Cybersecurity Culture. In particular, starting from the analysis of human nature and behaviour in relation to security, the book stresses the need for adopting a holistic approach to cybersecurity.

The first and foremost issue is to pay attention to the used language, since the development of cybersecurity awareness passes through a clear and understandable communication: We should consider that not all individuals working in organizations are IT or security experts, and we must not take anything for granted.

In this sense, we could talk of "social cybersecurity", not only for the role played by social media in criminal activities, but also for the role of communication in cybersecurity initiatives. In the end, people are the heart of every organization, regardless of their role.

If our focus is on technology all the time and not on humans together with technology, our approach to cybersecurity will be doomed to fail.

# Contents

# About the Author

**Isabella Corradini** is a social psychologist and criminologist. She is the Director of Themis Research Center (https://themiscrime.com/en/the-center), an interdisciplinary research organization for the psychological and social sciences, with a special focus on human factors in safety and security. She is also co-founder of the Link&Think Research Lab (http://link-and-think.blogspot.com/), which provides information, education and socio-technical analyses of information technologies. She is a recognized expert in the field of human factors in safety and security, and in the area of communication. She served as a Professor of Social Psychology at various Italian universities for 10 years. She is currently a Lecturer in master's and industrial training programmes and a consultant for national and international organizations on awareness initiatives aimed at developing safety and security cultures. A member of several technical and scientific committees, she is also the author of numerous articles, chapters and books on the above-mentioned topics, and the editor of a book series on the topic of reputation for a major Italian publisher.

# Chapter 1
# The Digital Landscape

**Abstract** Digital technologies play a significant role in our daily lives both professionally and personally. In particular, they have changed the way we communicate and do our work, and in general our social behaviour. Organizations are more and more dependent on information managed and exchanged through digital technologies; Internet of Things (IoT) and Artificial Intelligent (AI) applications continue to grow, producing a significant impact on our lives, and creating novel ethical and social issues to be faced. Not only common people and organizations use them, but also criminals exploit the advantages of technologies to enhance their modus operandi; while continuing to use traditional techniques, they are developing new expertise. The current landscape shows how cyberthreats continue to grow to the point where cyberattacks are included among the major risks to be concerned about for the next decade. Institutions and companies are improving their capabilities to strengthen cyber resilience, but results are far from being effective. In discussing what is possible to do to handle cybersecurity properly, we need—above all—to understand the relationship between people and technology, because people have to be considered as an essential part of any cybersecurity strategy. The serious impacts of cybercrime and the growth of cyberthreats have launched the theme of cybersecurity as a priority for nations, public institutions and private companies, with the aim of protecting their tangible and intangible assets. In this introductive chapter we are going to present the scenario, anticipating the main issues which will be discussed in this book.

## 1.1 Technology and Us

Over the last few years, cybersecurity has become one of the most important issues for organizations across all sectors, industries, and countries. Institutions and companies are improving their capabilities to strengthen cyber resilience, but results are far from being effective.

The current landscape shows how cyberthreats continue to grow to the point where cyberattacks are included among the major risks to be concerned about for the next decade (WEF 2019).

© Springer Nature Switzerland AG 2020

I. Corradini, *Building a Cybersecurity Culture in Organizations*,
Studies in Systems, Decision and Control 284,
https://doi.org/10.1007/978-3-030-43999-6_1

Before discussing cybersecurity and Cybersecurity Culture, we need to understand the relationship between people and digital technology, because the lack of a responsible use of technologies has to be tackled with a holistic approach to security.

It is a fact that technologies play a significant role in our daily lives both professionally and personally. In particular, they have changed the way we communicate and do our work, and in general our social behaviour. When we meet someone, for example, probably the first thing we do is to search his popularity on the Web. We can say that our expectations depend on technology more than human beings (Turkle 2011).

We cannot deny that modern digital devices promote a quicker communication and facilitate many tasks. Messaging apps are even replacing the use of emails to ensure the prompt delivery of messages.

The impact of digital technology on our lives is impressive: we are more and more dependent on it and most of us live in symbiosis with smartphones and social media. The crucial point is that people are becoming less and less autonomous in taking decisions and they are completely relying on technology. Some habits are lost, people live on smartphone and tablets, and "human contact" is now the real luxury goods.[1]

It is evident that the more we use sophisticated devices able to replace our actions, the more we will have difficulty in going back. Because of the possibility of owning a smartphone provided with facial recognition or fingerprinting, why should we use the old and common PIN? This is not "smart", in particular for the new generations who do not know the prehistoric applications of old phones. Even a PIN or a password can be risky, because everything depends on how people use them. The crucial point is not what we choose, but how aware we are of the risks when we use a certain application.

Digital transformation continues to spread exponentially and, in the future, we will be more and more interconnected; we can think, for example, of the development of Internet of Things (IoT), whose main characteristic is the fusion between the physical and the virtual world. The increasing number of objects connected to the Internet and their interactions, indeed, is going to lead to an ever-increasing interconnected ecosystem. They say that in the next years billions[2] of devices will be connected to the Internet and that they will be able to collect and exchange data through embedded sensors. They also say that thanks to this technology, an optimization of time and costs will be associated to a more productive work. However, they often forget to appropriately acknowledge security and privacy risks, deriving from the multiplication of data and the related need of managing it properly.

---

[1]https://www.nytimes.com/2019/03/23/sunday-review/human-contact-luxury-screens.html.

[2]Different estimates on the number of connected devices by 2025 are available. According to a forecast (2019) from International Data Corporation (IDC) in 2025 there will be 41.6 billion connected IoT devices, generating about 79.4 zettabytes of data. It is in fact evident that if connected devices grow, data produced raise too. https://www.idc.com/getdoc.jsp?containerId=prUS45213219.

Because of the significant impact on people's safety, security and privacy, the IoT threat landscape is rather complex (ENISA 2017)[3]; moreover, we already have evidence that it is possible for hackers to hijack connected vehicles.[4]

Besides IoT, the other relevant technological issue to be discussed is Artificial Intelligence (AI), whose applications already cover several industry sectors (Mc Kinsey Global Institute 2017), such as automotive, manufacturing, energy, transportation and financial services.

AI is not new technology, considering that this term was firstly used in 1950s (McCarthy et al. 1955), and that the creation of intelligent machines able to work and act like human beings has been of interest for several years. We still do not have a universally agreed definition of AI, since this depends on its classification, from research or business perspective (Del Ponte 2018).

Many are optimistic about the applications of AI, even converging with IoT, for instance in healthcare, where wearable sensors and mobile technology can improve the monitoring of the quality of life of patients. However, the debate on AI is very articulated, especially for the ethical and social implications for human beings. If on one hand, our lives might be improved by the use of this technology, on the other hand, the major risk is our complete loss of control.[5]

Any technological innovation which redefines processes and activities—as in the case of Artificial Intelligence—inevitably opens itself up to criticisms; the main concern is relative to what will happen if and when AI becomes better than humans in their cognitive tasks. Many operations, generally carried out by humans, will be more and more performed by technologies; on the contrary, we strongly believe that human beings cannot be completely replaced by machines.

After a few decades, ethical and social issues are becoming an essential aspect to be discussed and properly managed. From this perspective, the Ethics Guidelines for a Trustworthy Artificial Intelligence (European Commission 2019) represent a set of steps to promote an ethical AI development.[6] In fact, among the principles announced in the guidelines, it is said that AI should ensure adherence to ethical principles and values of human beings.

Another issue related to AI applications regards the responsibility by developers and the need of a specific normative (Martin 2019), because if some decisions can be

---

[3]ENISA is the European Union Agency for Network and Information Security, located in Athens. whose goal is to contribute to European cybersecurity policy and to support Member States and European Union stakeholders in the improvement of their capabilities. https://www.enisa.europa.eu/about-enisa.

[4]In 2015 security researchers Charlie Miller and Chris Valasek demonstrated how a car could be remotely hacked via its internet connection and paralyzed on a highway. https://www.wired.com/2015/07/hackers-remotely-kill-jeep-highway/.

[5]Elon Musk warned that AI can be an 'immortal' digital dictator able to rule mankind forever. https://www.cnbc.com/2018/04/06/elon-musk-warns-ai-could-create-immortal-dictator-in-documentary.html.

[6]According to this document, prepared by the High-Level Expert Group on AI, Trustworthy AI has three components: (1) it should be lawful, complying with all applicable laws and regulations (2) it should be ethical, ensuring adherence to ethical principles and values and (3) it should be robust, both from a technical and social perspective since, even with good intentions, AI systems can cause unintentional harm.

delegated to algorithms, someone must also be responsible for the possible negative consequences.

In the meantime, technologies are designed in order to be ever more persuasive. Already some years ago, scientists explained the mechanisms of persuasions by technologies on what we think and do (Fogg 2002).[7] Simplification, for example, can be considered a persuasive trigger to convince people: purchasing products on the web in one click is easier and more comfortable than going to a physical shop. In addition, personalized products and services are designed to make people feel like special buyers; so, individuals tend to purchase online (Zhen et al. 2017) for emotional attachment, uniqueness seeking, identity expression, and so on.

Simplification of our lives seems to be the priority for businesses, but the question is whether all these technologies are really an advantage for us. The anxious search for improving people's life and the "folly of technological solutionism"[8] (Morozov 2013) risk to be a strong limit on human's creativity and freedom.

We should reflect on what happened when email was introduced to public use. We were convinced that Internet and emails would have simplified our lives and that we would have gained more free time. This is partially true, but it is equally true that we are now oversaturated with emails and messages, suffering from an information overload, dangerous to our health.

We are living in the digital era, we cannot and must not escape from it, but we need to rebuild a new balance with technologies to preserve our personal and social identity.

## 1.2 Everything is Cyber

"Cyber" is everywhere. This neologism is widely used to describe everything: crimes, jobs, places, fields of application, even TV series. In everyday language, we say cyberbullying, cybercrime, cyber experts, as if these terms were something completely new. However, these terms—without the prefix cyber—refer to concepts we already know in the physical world.

Cyber prefix derived in the late 1940 from the field of cybernetics, the study of communication and control systems in living beings and machines.[9] Its use has grown over the last few years; often, it has nothing to do with the classic concept of cybernetics.

The current attention is especially focused on the negative aspect of cyberspace, perceived as "an evolving realm of human interaction with specific security and

---

[7]Fogg, director of the Behavior Design Lab at Stanford University, coined the term "Captology" to indicate the study of computers as persuasive technologies.

[8]In his book "To save Everything, Click Here: The Folly of Technological Solutionism" the author, a critical voice on the social role of the Internet, identifies two significant issues: solutionism (the idea that technology can solve all of mankind's problems), and Internet-centrism (a sort of ideology exaggerating the capability of Internet as a stable force in our lives).

[9]The word cybernetics was first used by the American mathematician Norbert Wiener in his book published in 1948 (Wiener 1948).

**Table 1.1** Definitions of cybersecurity in online dictionaries

| |
|---|
| Measures taken to protect a computer or computer system (as on the Internet) against unauthorized access or attack (Merriam-Webster) https://www.merriam-webster.com/dictionary/cybersecurity |
| Preventative methods used to protect information from being stolen, compromised or attacked. It requires an understanding of potential information threats, such as viruses and other malicious code. Cybersecurity strategies include identity management, risk management and incident management. (Techopedia) https://www.techopedia.com/definition/24747/cybersecurity |
| The state of being protected against the criminal or unauthorized use of electronic data, or the measures taken to achieve this (Lexico Dictionaries) https://www.lexico.com/en/definition/cybersecurity |
| Precautions taken to guard against crime that involves the Internet, especially unauthorized access to computer systems and data connected to the Internet. The state of being protected against such crime. (Dictionary) https://www.dictionary.com/browse/cybersecurity |

defence concerns" (McGuffin and Mitchell 2014). It is a fact that real wars are fought in cyberspace, by now recognized as the fifth domain of warfare, after land, sea, air and space (NATO 2016). The increase of cyberthreats has launched the theme of cybersecurity as a priority for nations, institutions and companies, with the aim of protecting their tangible and intangible assets.

In the "cyber-words" landscape, we observe an interchangeable use of the different terms. For IT specialists these words are familiar, but not everyone is an expert in this field. Actually, as other fashionable terms, there is still little understanding of the real meaning of cybersecurity (Schatz et al. 2017).

The foremost issue is therefore to define a correct approach to the use of cyber-terms, developing a common language to be shared, and above all, understandable to everyone. Despite many papers and books focusing on this issue, there is not a universally agreed-upon definition of cybersecurity, or how to write the word (cybersecurity or cyber security or cyber-security).[10]

After a quick search on the Internet, we can find several definitions deriving from available dictionaries; some of them are reported in the following table (Table 1.1).

Among other definitions, one of the best known defines cybersecurity as (NIST, Glossary, 2013).

> the ability to protect or defend the use of cyberspace from cyberattacks.

Considering the extension of cyberspace and the multiplicity of actors and activities involved, it is clear that the focus of cybersecurity is on the protection of computer systems (computers, smartphones, telecommunications networks, etc.) and on data in electronic form, both from external and internal attacks.

A wider characterization, including goals, actions and activities, is given by ITU (International Telecommunication Union 2008), which defines cybersecurity as

---

[10]In literature both versions are used. According to Professor of Linguistics Suzanne Kemmer, cybersecurity should be one word because cyber is not a free-standing word https://eforensicsmag.com/dr-shem/.

> the collection of tools, policies, security concepts, security safeguards, guidelines, risk management approaches, actions, training, best practices, assurance and technologies that can be used to protect the cyber environment and organization and user's assets.
>
> Organization and user's assets include connected computing devices, personnel, infrastructure, applications, services, telecommunications systems, and the totality of transmitted and/or stored information in the cyber environment.
>
> Cybersecurity strives to ensure the attainment and maintenance of the security properties of the organization and user's assets against relevant security risks in the cyber environment. The general security objectives comprise the following: Availability, Integrity, which may include authenticity and non-repudiation, and Confidentiality.

From the analysis of the definitions listed above, "protection" is the keyword which emerges, helping us to understand the goals of cybersecurity.

Nevertheless, we must not take it for granted, because actually many people confuse the various terms, using them interchangeably, for example cybercrime with cyberbullying, or with cybersecurity. Even though everything is cyber, not everything means the same thing.

Cybersecurity conferences are typically attended by specialists; even though this kind of debate is always useful and stimulating, the real challenge is to involve inexperienced people, in order to open them up to the subject. In fact, common people use digital technologies, often without being aware of the risks they are exposed to. This careless use includes personal and professional lives, given that the two are not clearly separated. Moreover, when individuals use digital technologies unwisely, it is difficult to think that their behaviour will be any different at work. Hence, the need to improve awareness and education in cybersecurity.

Over the last few years, the term cybersecurity has gradually replaced other terms previously used, like "Computer Security", "IT Security" or "Information Security". Moreover, information security and cybersecurity are often used interchangeably.

Actually, Information Security (or InfoSec) refers to (NIST, Glossary, 2013)

> the protection of information and information systems from unauthorized access... to provide confidentiality, integrity, and availability.

Its goal is the protection of data in any form, not necessarily electronic data; therefore, its meaning should be broader than cybersecurity. In the debate, some authors consider cybersecurity and information security differently (e.g. Von Solms and Van Niekerk 2013), arguing that cybersecurity goes beyond the boundaries of Information Security.

Other differences are discussed between Information Security Culture and Cybersecurity Culture (Reid and van Niekerk 2014) on the basis of the context: the former (Information Security Culture) would be cultivated in organizational context (a controlled environment), while the latter (Cybersecurity Culture) would appear to concern a societal context (less controllable). However, we have to consider that people working in organizations are part of the societal context too, and it is therefore difficult to neatly separate the two areas.

Another significative element is the frequent use of the term cybersecurity compared to the others cited above. Both in media communication and in common language, cybersecurity is certainly more fashionable and attractive to the audience rather than information security; for commercial purposes, organizations and institutions prefer to use cybersecurity. After all (Wamala 2011).

> cybersecurity is information security with jurisdictional uncertainty and attribution issues.

In this book we have decided to use cybersecurity—also referring to the protection of information—for several reasons:

- the protection of information is an important issue for cybersecurity;
- because of digital transformation, information is going to be more and more digitalized;
- the word is familiar to many people, which can help the implementation of security awareness programmes.

Whether cyber or not, the heart of this book is represented by people: we cannot think of handling cybersecurity problems if we neglect the role of human beings. The challenge is to transform individuals from weak to strong links of security.

## 1.3 The Digitized Crime

Besides benefits produced by the use of digital technologies, there is also a "dark side" represented by criminal activities developed in the digital environment. Research has shown how Internet facilitates crime and deviance (Stalans and Finn 2016), providing many opportunities to criminals.

With the digitization of society, therefore, also crime has become digital (Leukfeldt 2017). On the one hand, new offenses have been produced (e.g. infecting computers with malware); on the other hand, traditional forms of crime exploit information technology for their realization (e.g. internet fraud and cyberstalking).

The concept of opportunity is well-known in criminology, since it plays an important role in causing crime (Felson and Clarke 1998) and affects motivations driving criminal actions. This notion was initially studied with respect to predatory crimes, such as robbery, theft, burglary, whose commission is strictly connected to economic gain. The approach of "crime opportunity" gives much relevance to situational elements to explain the occurrence of crime. In fact, criminal dispositions are seen in relationship with opportunities perceived and acted on by criminals.

Most important theories relating to this approach are the "routine activity theory" (Cohen and Felson 1979; Felson 2002), and the "rational choice theory" (Cornish and Clarke 1986; Clarke and Felson 1993).

The routine activity theory is focused on the characteristics of crime and stresses the role of environment. In particular, the convergence of three elements is considered essential for the realization of a crime: a potential offender, a suitable target, and the absence of a capable guardian, where the term does not necessarily denote a policeman, but any person that, by only her presence, can prevent crimes.

The rational choice theory, inspired by the work of Cesare Beccaria in the 1700's, starts from the concept that individuals freely choose their behaviour, being conscious of their actions, and therefore they are responsible for their choices. In fact, in deciding their actions, they evaluate costs and benefits.

Crime opportunity approach offers some interesting insights for the study of cybercrime, too. In accordance with research (e.g. Hutchings and Hayes 2009; Pratt et al. 2010; Lastdrager 2014) the routine activity theory can be applied to the digital world, since, for example, time spent on the Internet and on social media is a routine activity which takes part in the victimisation process.

In addition, social and technological changes are also included among the principles considered in the approach.[11] Indeed, digital technologies have produced significant changes in people's habits and behaviour, studied and exploited by cybercriminals for their illegal actions.

Regarding the involvement of IT, it is preferable to distinguish (Leukfeldt 2017) cyber-dependent crimes and cyber-enabled crimes. In the former, crimes are committed through the use of IT, e.g. hacking; in the latter, IT is an enabling factor, e.g. a fraud on the internet. Both types are included in cybercrime.

The area of crime on the Internet, however, covers a large spectrum and evolves experientially (Gordon and Ford 2006); hence, it might appear more functional defining categorical terms rather than describing cybercrime precisely (Finnie et al. 2010). Moreover, different cyber threat actors move on the Internet with different motivations and means (Ablon 2018).

The use of the prefix cyber can lead people to confuse the different phenomena, to the point where cyberbullying and cybercrime are used as if they were the same thing, alongside cybercrime and cyberterrorism.

To clarify the meaning of these concepts, we can use another perspective of classification, which considers the different targets together with motivation and methods:

- cybercrime, when the target is a property (like, e.g., in a fraud through the Internet)
- online harassment, attacking an individual (as in the cases of cyberstalking and cyberbullying)
- cyberterrorism and cyberwarfare, when the target is a country and its representative institutions.

Despite having the digital context as a common denominator, these phenomena involve various actors, motivations and dynamics. Considering the wide variety of literature, and referring to the type of classification listed above, here we describe

[11]According to Felson and Clarke (1998) there are ten principles of opportunity and crime: 1. Opportunities play a role in causing all crime; 2. Crime opportunities are highly specific; 3. Crime opportunities are concentrated in time and space; 4. Crime opportunities depend on everyday movements; 5. One crime produces opportunities for another; 6. Some products offer more tempting crime opportunities 7. Social and technological changes produce new crime opportunities 8. Opportunities for crime can be reduced 9. Reducing opportunities does not usually displace crime 10. Focused opportunity reduction can produce wider declines in crime.

briefly the concepts referring to the different crime activities developed in the digital context.

### 1.3.1 Cybercrime

This is a wide area where cybercriminals are essentially motivated by financial gain: they are interested in stealing data and in everything that can be monetized on the cybercrime black market.

Cybercrime continues to mature and to become bolder and bolder (Europol 2019); the availability of digital technologies has provided cybercriminals with more opportunities, promoting a combination between traditional and innovative techniques. The real issue is if we can consider cybercriminals as a new type of offender or as a traditional one. It is not simple to provide an answer to this question.

On the one side there are many different types of cybercrime, based on the combinations in different degrees of traditional and digital techniques, and they are continuously evolving. On the other side, the study of this kind of offender has severe limitations, since many factors have to be considered: demographic variables, personality traits, personal motivation, technical skills, social habits, family characteristics, type of crime.

From this point of view, several studies have focused on hackers' profiling, through the analysis of their technical skills and motivations. According to research, it is possible to identify various kinds of professional hackers (Chiesa et al. 2009; Seebruck 2015) and different motives (Madarie 2017), which include—among others—technical challenge, peer recognition, testing system security and tangible rewards.

The image itself of hackers has evolved over time: considering the complexity and the gravity of cyberattacks, now it is rather difficult to image them as young people working alone in isolated basements. We therefore need to go beyond certain stereotypes. There are certainly young individuals with high technical skills motivated by personal challenges and by the desire of discovering system weaknesses, such as in the case of "ethical hackers" (or "white hats"). However, when we talk about organized crime, a large number of professionals with malicious intent, so called "black hats", is involved. Other typologies of hacker include: hackers who protest on the Internet for social causes ("hacktivists"); hackers placed between white and black hackers ("grey hats") because of the combination of opposite intentions (good or bad); hackers employed by governments to acquire information on foreign countries ("state sponsored hackers"); hackers who want to demonstrate high capacities but, in reality, own little technical skills ("script kiddies").

Classifications and profiling are never easy and sometimes can even be misguided. Human nature is too complicated and unpredictable to be closed in rigid schemes. However, the comparison between online and offline characteristics of offenders is of particular interest for criminology. This analysis, in fact, can be useful to shape adequate policy in terms of detection, intervention and punishment (Hargreaves and Prince 2013).

In this sense, the comparison between traditional and cyber-crime can provide useful elements in order to better identify characteristics of cybercriminals. Generally speaking, cybercrime is a type of crime which can count on the use of the current digital technologies and, differently from the traditional forms, it takes place in the digital context, overcoming physical and temporal spaces. Therefore, attacks can reach more than one target at once, with more benefits and less risks. Moreover, lack of physical contact in cybercrime facilitates the completion of the illegal activity, affecting psychological dynamics within cyberspace: moral disengagement is typically associated to a different perception of risk both in criminals and in victims (Sect. 2.3), and in general, we observe an altered public perception of cybercrime compared to violent crimes.

We have extensively underlined the significant role of IT technology in cybercrime; currently, these technologies are an essential support for the execution of traditional forms of crime, too.

The term cybercrime has many different meanings and is used to refer to illegal activities of diverse nature. It can include offences against the confidentiality, integrity and availability of computer data and systems (Council of Europe's Budapest Convention on Cybercrime 2001), but also offences committed by means of computer systems. With this extensive approach cybercrime covers all those forms of crime using computers, such as fraud and online harassment. While there is a high consensus on the broad areas of criminalization, many differences depend on national approaches to criminalization of cybercrime actions (UNODC—United Nations Office on Drugs and Crime 2013).

The key element is that all these types of activities involve a computer and a network, and in general all networked devices, in agreement with the evolution of digital technologies. Therefore, if in the previous definitions of cybercrime, the focus was mainly on computers, now we need to consider the wide use of smartphones; therefore, talking about computing devices or networked devices is more appropriate.

Cybercrime is a serious issue for people and organizations, causing significant economical damages. In ransomware attacks, for example, attackers, after gaining access to the victim's system, encrypts her documents and files; the system is blocked until the ransom is paid, usually in bitcoins. Data and information are an integral part of people's business and private life, and it is therefore easy to imagine why ransomware can be profitable for attackers.

Analysing the rise of technological applications, it is possible to anticipate new trends: since mobile banking has been growing and surpassing online banking, we can expect the increase of mobile malware targeting users of this service (Europol 2018).

Finally, it is hard to separate cybercrime from cybersecurity: they are mutually reinforcing even though they follow a different rational[12]: cybercrime strategy is based on the rule of laws, criminal justice and human rights, while cybersecurity strategy aims at enhancing security, trust, resilience and reliability of ICT.

### *1.3.2 Online Harassment*

We consider online harassment as a separate class for at least two reasons. The first is that in cybercrime the economical motivation is dominant, while in harassment, such as cyberstalking and cyberbullying, perpetrators are more interested in exerting a strict control on their targets and harming them. The second reason is connected to the different entity of consequences, considering that persistent harassment experiences can produce significant physical and mental effects on victims.

Cyberstalking and cyberbullying are the most popular forms of harassment, where individuals are attacked or threatened or intimidated through the use of electronic means. Both types of harassment have a relationship with cybercrime, since some of its actions can also be used with the intent of harming targets, such as identity theft, attack on data (sending virus), trolling,[13] use of social engineering tactics to gather information on victims.

Harassing behaviour is committed repeatedly in order to humiliate, persecute and control. Among the activities developed to hurt, we can find: posting unwanted pictures or information on social media, sending negative messages via text or emails, sharing private information to cause embarrassment.

In most cases victims of cyberbullying are young people and school-aged children; however, the definition of cyberbullying can be related to different perspectives and much research is available on this theme (e.g. Campbell 2005; Patchin and Hinduja 2006; Shariff and Hoff 2007).

In cyberstalking a perpetrator uses electronic communication devices to repetitively harass another person, through control and intimidation (Spitzberg and Hoobler 2002). Internet is used by stalkers to find information on potential victims and monitor them; comparing stalking and cyberstalking, the virtual world may more easily induce perpetrators to act than offline environments (Dreßing et al. 2014).

Regardless of the parties involved (minors or adults), it is possible to observe some similarities between cyberbullying and cyberstalking; for instance, the continued focus on a specific person and the constant provocation generates for victims a situation difficult to handle; in fact, they feel they are in a subordinate position and for them it is hard to react adequately.

---

[12]PGG (2018) Cybercrime@EAP Project, Cybercrime and cybersecurity strategies in the Eastern Partnership region. Results of series of national workshops Eastern Partnership countries. This project is funded by the European Union and the Council of Europe. https://rm.coe.int/eap-cybercrime-and-cybersecurity-strategies/168093b89c.

[13]Trolling is the act of leaving an insulting message on the internet in order to annoy someone (Cambridge Dictionary) https://dictionary.cambridge.org/it/dizionario/inglese/trolling.

Moreover, cyberbullying can take on different forms, including cyberstalking (Willard 2007). Other types are: denigration (arguments between protagonists), harassment (repeated ongoing assault), flaming (cruel speech), impersonation (identifying as the target), outing and trickery (spreading information or images embarrassing for the target), and exclusion (marginalisation of the target).

### *1.3.3 Cyberterrorism and Cyberwarfare*

Literature on cyberterrorism is extensive and specialized, but its definition is not so linear. The convergence between terrorism and cyberspace (Denning 2000) is what best helps to understand the phenomenon. With respect to the traditional form, cyberterrorism has significant peculiarities, determined by the virtual environment (Gordon and Ford 2004), which still poses interesting questions about psychological effects of the actions conducted by different perpetrators. Indeed, in cyberterrorism, the goal is not necessarily to kill someone, but to cause significant disruption to a community (Shipley and Bowker 2014).

We are fully aware that such a complex issue requires a broader discussion, also considering the difficulty of its definition, but here the goal is to distinguish the different criminal activities which take place in cyberspace, in order to avoid confusion between them. Cyberterrorism can be defined as (NATO 2008):

> cyber-attack using or exploiting computer or communication networks to cause sufficient destruction to generate fear or intimidate a society into an ideological goal.

Given the many different political, ideological and religious goals, a large number of potential targets have to be considered: from authorities, to national critical infrastructures,[14] to representative entities.

It is however important to analyse the relationship between cyberterrorism and cybercrime, since vulnerabilities in networks and systems can be used for both, even if for different objectives. Hence, hacking, viruses, malware or manipulation of data are part of modus operandi of cyberterrorists. It is therefore not easy to distinguish a cyberattack realized by terrorists or by hackers (Pladna 2008).

Cyberwar (or cyberwarfare) refers to military operations fought within cyberspace; hence, all the activities are carried out with specific political-military purposes.[15] We usually talk of state-sponsored actors, whose task is to penetrate and achieve top-secret information of other governments. For this goal, they receive funds and assistance from a nation state to develop its particular interests.

---

[14]Critical infrastructures provide essential services for the maintenance of vital societal functions, such as health, safety, security, economic and social well-being of people; their disruption or destruction would have negative consequences for the security of countries and its citizens (see, for example, Council Directive, 2008/114/EC).

[15]For definitions and details see, for example, the Tallin Manual, the most comprehensive analysis on how existing international law applied to cyberspace. https://ccdcoe.org/research/tallinn-manual/.

Highly experienced hackers and sophisticated techniques characterize cyberwarfare. For instance, APT (Advanced Persistent Threat)[16] attacks target significant institutions belonging to the sector of national defence and companies holding highly sensitive information. Here, the goal is to gain government data, military plans and intellectual property.

Moreover, both for cyberterrorism and cyberwarfare we must not forget the important role of propaganda, as discussed in Sect. 1.5.

In conclusion, we should not forget that whether they are civilians, military or financial organizations, everybody's life depends on digital infrastructures; therefore, cyberterrorism and cyberwar take place on a complex field to be defended.

### *1.3.4 Summary*

As discussed in this paragraph, the area of crime on the Internet involves a plethora of illegal and criminal activities and different targets, summarized in Table 1.2.

Most of the illegal actions described above are well known in the physical world. Thefts, fraud, extortions, and terrorist attacks, all occurred before digital revolution, but now they can rely on the means offered by the digital context, including data and information stored in digital devices.

In short, we need to be aware of the risks on the Internet, since we live in the digital world and we could all be possible targets.

**Table 1.2** The area of Internet crime

| Type of crime | Targets | Motivation | Methods (examples) |
|---|---|---|---|
| Cybercrime | Anyone (individuals and organizations) | Financial gain | Identity theft and fraud, cyberespionage, cyberextortion |
| Online harassment | Individuals, groups | Intimidation, denigration, assault | False accusations, identity theft, monitoring target's online activities, trolling, sending virus |
| Cyberterrorism and cyberwarfare | Authorities, critical infrastructures | Ideological and political reasons | Viruses, hacking, propaganda |

[16]An Advanced Persistent Threat (APT) is a prolonged targeted cyberattack where an intruder gains access to a network and remains undetected for a certain period of time. This type of attack is based on sequential stages, including—among others—gaining access, moving laterally, taking data and covering their tracks https://searchsecurity.techtarget.com/definition/advanced-persistent-threat-APT.

## 1.4 Data is Business

In the digital era, data is a valuable asset both for individuals and for companies, and it represents the new currency of the digital economy. As Meglena Kuneva (European Consumer Commissioner) said in her speech in 2009, "Personal data is the new oil of the internet and the new currency of the digital world".[17]

With the increasing digitization of society, the amount of data has grown, too. It has been generating a rich field of opportunity for economic and societal value creation (World Economic Forum 2011). At the same time data has acquired a new and attractive meaning, since it is produced by people and concerns people's attitude, preference and behaviour. In short, data is the records of people's lives.

It is not a secret that companies are more and more interested in capturing and storing consumers' data for analysing their behaviour and defining their best marketing strategy. However, data is business even for cybercriminals, who steal commercial data and exploit the amount of information about users' behaviour available on the Internet for planning their activities.

On the other hand, how can we imagine spending our days without emails, social media, and smartphone? It would be an excellent exercise (we should try to do it), but if we are immersed in a digital world we are forced to follow its mechanisms. Most people are convinced that they have complete control of their digital environment and of data, but we know that this is not true.

Hence, from a criminal perspective, data represents a key commodity, which can be used for several purposes, such as immediate financial gain, complex fraud and extortion (Europol 2016).

Companies are an appealing target to criminals, since they hold massive amount of information; consequently, we can imagine how profitable a data breach at a large company is. At the same time, data is also a valuable resource for investigations by law enforcement agencies (Europol 2019).

Cybercrime is a productive business model and typically associated with low risks, especially when compared to traditional illegal activities. The so called "crime-as-a-service" (CaaS) provides the most varied services and skills, such as: malware (malicious software), botnets (networks of computers infected with malware), hacking services, databases of stolen personal information, penetration testing of potential targets,[18] and much more. Considering the different nature of services, it is clear that for many types of cybercrime, a high degree of organization and specialization is needed (Broadhurst et al. 2014).

---

[17]http://europa.eu/rapid/press-release_SPEECH-09-156_en.htm.

[18]Penetration testing simulates a malicious attack to verify how easy is to access networks or systems in order to steal valuable data. Previously, this practice was mainly applied with government organizations; now, it is also used in many businesses.

A prosperous market for data business is the Dark Web, the segment of the Deep Web widely used for illegal activities, appreciated by criminals for anonymity and multiple level of encryption guaranteed by special software.[19]

In the Dark Web, drugs, weapons, malware, and credentials are marketed[20]; criminal groups and terrorists are very active in this part of the web. We must not think that all the Dark Web is populated by criminals; in fact, it is also used by people who need to share sensitive information, because of the censorship, and by companies interested in monitoring stolen corporate account information. Here too, considering the multiplicity of activities developed in this area of the Web, law enforcement and governmental agencies are interested in monitoring them.

The main issue is: how much is data worth?

Searching on the Internet with queries like "the cost of data on the web" or "how much is your personal data worth to hackers?", we can find various answers and estimates.

First of all, there is not a simple answer and a full answer is even more complicated (Ellis 2019). The value typically depends on the types of data and its potential yield: medical data, for example, is considered highly valuable to hackers because it contains a lot of sensitive information, which can be used, for example, for medical fraud. The Anthem medical data breach is one of the major intrusions into a U.S. healthcare company, occurred in 2015. The hackers gained access to up to 80 million records, including full names, physical addresses, emails, medical IDs, income data, etc.[21]

However, it is not clear to identify the determinant elements for pricing trends within the cybercrime economy, since prices can change drastically (Report Flashpoint 2019). Amounts can change according to other information, like users' age, how common information is, the country individuals live in, and so on. Prices change year by year, too.

---

[19]Graphically, Internet is often represented like an iceberg. At the top there is the visible and free surface, that can be reached through search engine; below, there is the Deep Web, an area not accessible without having a direct address (as it happens for services like Google Drive and Dropbox or for dynamically created pages). Within the Deep web, there is the Dark Web, whose addresses are masked and whose pages can be accessed only through special software, such as TOR or I2P, guaranteeing anonymity both to users and servers.

[20]According to ImmuniWeb researchers (2019), more than 21 million credentials belonging to Fortune 500 companies have been object of trading between cybercriminals. In particular, 95% of these contained unencrypted, or brute forced and cracked by attackers, plaintext passwords. Many corporate credentials are used for spear-phishing and password re-use attacks against the largest global companies. https://www.immuniweb.com/blog/stolen-credentials-dark-web-fortune-500.html.

[21]The breach was discovered in January 2015, but in December 2014 Anthem employees had already noticed some suspicious activities https://resources.infosecinstitute.com/category/healthcare-information-security/healthcare-attack-statistics-and-case-studies/case-study-health-insurer-anthem/#gref.

Moreover, even for the same product different prices are available. In the cybercrime black market, stolen credit cards can have different values: high or no-limit cards or those who are equipped with chip and PIN are more valuable compared to others (Ablon 2018).

A widely used means to steal data is phishing email, whose goal is to retrieve user data by misleading the receiver to execute an unwanted action such as to complete a data form or to update passwords for accessing a bank account, and so on. These messages often contain elements of urgency, in order to lead people to act immediately. Phishing campaigns are also used to infiltrate organization networks by sending fraudulent emails to employees. The hasty management of these emails is one of the major strategies used by criminals and related to security incidents and data breaches (Verizon 2018).[22]

Additionally, there are also emails sent to targeted recipients on the basis of accurate analyses, as in the case of spear phishing, which uses social engineering techniques (Chap. 3). These tactics are often used against company personnel in order to get information for industrial espionage purposes.

To protect their business, companies and institutions must monitor the underlying threats, paying attention to their evolution, assessing their cyber risk profile and implementing all technical and organizational measures needed.

Preparedness of employees is one of the most important measures to ensure data protection; therefore, as discussed in this book, it is imperative for companies and institutions to realize adequate cybersecurity training programmes.

## 1.5 The Role of Social Media

In the last few years the role of social media platforms has rapidly grown. They are an important tool of communication, used for different purposes, from commercial activities and public debate to raising funds and supporting emergency situations.

The number of people using internet and social media has been growing year after year.[23] These platforms satisfy several human needs; in fact, by using them people can describe themselves, share information and experiences, and be informed in real time.

Even companies take advantage of using social media; indeed, they can promote their products, attract new customers, and create communities around a certain brand. These activities are made possible by the creation of public profiles and the articulation of contact lists (Boyd and Ellison 2007).

---

[22]According to Verizon reports an incident is "a security event that compromises the integrity, confidentiality or availability of an information asset"; instead a breach is "an incident that results in the confirmed disclosure—not just the potential exposure—of data to an unauthorized party.".

[23]See, for example, the analysis conducted every year by the socially-led creative agency "We are social" in cooperation with "Hootsuite", the popular platform of social media management.

Social media has almost replaced traditional media, changing the paradigms of communication. Just think of the "political narrative", where political actors become promoter of themselves, constantly in touch with their audience. These platforms facilitate the fast spreading of information, overcoming any geographical and physical border.

These features are exploited for developing illegal activities, too. For example, criminals use social media platforms to spread malware, buy and sell hacking tools and expertise, to the point where the amount of these activities into the "Web of Profit"[24] is growing year by year.

Social media has also become a fertile environment for hate speech and fake news. In fact, hiding behind a screen offers comfortability and protection to individuals, who feel encouraged to use hostile behaviours against others, avoiding—at the same time—a face to face confrontation.

Hate speech is a growing problem capable of producing direct and indirect effects on individual's psychological well-being (SELMA 2019) and, in general, on society. There is not a universally accepted definition of hate speech, given that it covers many types of hateful and harmful expressions (e.g. Brown 2017).

A broad and update definition of hate speech is given by ECRI[25] in General Policy Recommendation (2016) which defines hate speech:

> the advocacy, promotion or incitement, in any form, of the denigration, hatred or vilification of a person or group of persons, as well as any harassment, insult, negative stereotyping, stigmatisation or threat in respect of such a person or group of persons and the justification of all the preceding types of expression, on the ground of 'race', colour, descent, national or ethnic origin, age, disability, language, religion or belief, sex, gender, gender identity, sexual orientation and other personal characteristics or status.

Handling hate speech requires balancing the freedom of expression with the need of fighting hatred and violence. Therefore, beyond the adoption of codes and specific regulations, public awareness on this subject is one of the most effective antidotes to the problem.

Fake news is another important issue to deal with. This term became very popular in November 2016 during the US presidential elections. It has been estimated that the frequency in Google Trends of this term after this event increased very sharply (Martens et al. 2018). Fake news is not a new topic, because rumours and false information were also widespread in the past. Here too, what makes the difference is the digital context, because of its power of facilitating the diffusion of information. The effects can have a significant impact, considering that fake news can alter the public debate, influence the decision-making process and manipulate the media agenda (Marwick and Lewis 2017).

---

[24]According to the study "Social Media Platforms and the Cybercrime Economy" conducted by Bromium in cooperation with University of Surrey (Dr. Mike McGuire), social media platforms are enabling cybercrime, making easier distributing malware. Cybercriminals are earning at least $3.25 bn per year from social media-enabled crime. An important highlight of the report is that one in five organizations has been attacked by malware delivered by means of social media. https://www.bromium.com/social-media-platforms-cybercrime-economy/.

[25]European Commission against Racism and Intolerance (ECRI).

The concern about the effects of fake news for democracy is a major topic for nations, since false information and social media can be used to harm the reputation of specific political targets and change the outcome of elections. Actually, consequences can be more devastating and threaten the stability of a country. On the other hand, in a broader perspective, we are aware that social media is an effective tool for propaganda and disinformation in cyberwarfare (NATO 2016).

The intention to cause harm and falseness define three forms of fake news (Wardle and Deraskan 2017). So, it is possible to spread false information in an accidental way and without causing harm (misinformation); using an intentional way to cause harm (disinformation); finally, it is also possible to share genuine information and cause harm (malinformation).

Fake news diffusion often exploits specific programs able to react automatically to what happens in the given medium, the so-called bots. Whether it is done by people or bots, the act of sharing gives credibility to the news, and induces people to trust instead of investigating the original source (Susarla 2018), and sharing again. In this way, fake news becomes viral and can affect people's emotion; when fear and anger are stimulated, it may happen that individuals or groups react aggressively.

Despite online aggressions being mostly confined to the space of social media, we cannot exclude that provocative messages—shared by multiple individuals and groups—lead people to an explosion of collective anger, like street protests and aggressions.

Cognitive and social factors are involved in the explanation of how fake news works and why it can be successful. In particular, social media facilitate people's choice according to their beliefs, as sharing their favourite readings (Flaxman et al. 2016). In fact, "echo-chambers" are based on a confirmation bias: among the available information, people tend to select information confirming their thoughts, while ignoring what disconfirms them. If cognitive biases are part of human nature, limitations of the cognitive analysis can produce serious consequences.

The paradox is that in the Information Age it is very difficult to be able to handle the large amount of information, as well as to distinguish good info from bad. Too much information available on the Internet generates what has been called "information pollution" (Nielsen 2003), where the mass of irrelevant information hampers us to find what we really need. The challenge for individuals is to be in-tune with the "stream" of contents (Boyd 2010), becoming aware of the information flow in order to grab the most relevant. Moreover, in this *mare magnum* it is easy that true information and false facts contaminate each other.

The increasing use of Artificial Intelligence, if on one hand can help to fight fake news, on the other hand can strengthen its creation. In particular, thanks to the use of machine learning algorithms, it is possible to create "deepfakes", constructing human faces resembling real people and imitating their voices. Celebrities are the most affected ones, but everyone is a potential target, as long as enough digital images and data are available.

This issue also regards enterprises, since deepfake voices fraud can be used to trick companies into sending money to a fraudulent account.[26] Finally, this is a critical issue for national security given that sophisticated fake videos can be created and spread by terrorists for their goals.

It will be more and more difficult to understand when a video is computer-generated or not. Here too, the human factor plays an important role, given that not everyone takes time to verify if a piece of news is accurate or not. People aware of the consequences should always verify the information before sharing it.

## References

Ablon, L.: Data Thieves, Testimony before the House Financial Services Committee, Subcommittee on Terrorism and Illicit Finance (March 2018)

Boyd, D., Ellison, N.: Social network sites: Definition, history, and scholarship. J. Comput. Mediat. Commun. **13**, 210–230 (2007)

Boyd, D.: Streams of content, limited attention: the flow of information through social media. Educause Rev. **45**(5), 26–36 (2010)

Broadhurst, R., Grabosky, P., Alazab, M., Chon, S.: Organizations and cyber crime: an analysis of the nature of groups engaged in cyber crime'. Int. J. Cyber Criminol. **8**(1), 1–20 (2014)

Brown, A.: What is hate speech? Part 2: family resemblances. Law Philos. **36**(5), 561–613 (2017)

Campbell, M.A.: Cyber bullying: an old problem in a new guise? Aust. J. Guid. Couns. **15**, 68–76 (2005)

Chiesa, R., Ducci, S., Ciappi, S.: Profiling hackers: the science of criminal profiling as applied to the world of hacking. Auerbach Publications (2009)

Clarke, R.V., Felson, M.: Routine Activity and Rational Choice. Advances in Criminological Theory, vol 5. Transaction Books, New Brunswick, NJ (1993)

Cornish, D., Clarke, R.V.: Introduction. In: Cornish, D., Clarke, R. (eds.) The Reasoning Criminal. New York, Springer, pp. 1–16 (1986)

Council of Europe: ETS 185, Convention on Cybercrime (23.11.2001)

Cohen, L.E., Felson, M.: Social change and crime rate trends: a routine activity approach. Am. Sociol. Rev. **44**, 588–608 (1979)

Denning, D.: Cyberterrorism, testimony before the special oversight panel of terrorism committee on armed services, US House of Representatives, 23 May (2000)

Del Ponte, L.: European Artificial Intelligence Leadership, The Path for an Integrated Vision. Policy Department for Economic, Scientific and Quality of Life Policies, European Parliament, Brussels (2018)

Dreßing, H., Bailer, J., Anders, A., Wagner, H., Gallas, C.: Cyberstalking in a large sample of social network users: prevalence, characteristics, and impact upon victims. Cyberpsychology Behav. Soc. Netw. **17**, 61–67 (2014)

Ellis, W.J.: How Much Does Your Data Cost on the Dark Web?–We Checked (2 June 2019). https://privacyaustralia.net/dark-web-personal-data/

ENISA: Baseline Security. Recommendations for IoT in the context of Critical Information Infrastructures (November 2017). https://www.enisa.europa.eu/publications/baseline-security-recommendations-for-iot

[26]https://www.theverge.com/2019/9/5/20851248/deepfakes-ai-fake-audio-phone-calls-thieves-trick-companies-stealing-money.

European Commission against Racism and Intolerance, ECRI General Policy Recommendation N.15 on Combating Hate Speech (2016). https://www.coe.int/en/web/european-commission-against-racism-and-intolerance/recommendation-no.15
European Commission: Ethics Guidelines for Trustworthy AI (2019). https://ec.europa.eu/digital-single-market/en/news/ethics-guidelines-trustworthy-ai
Europol: The Internet Organised Crime Threat Assessment (IOCTA) (2016). https://www.europol.europa.eu/activities-services/main-reports/internet-organised-crime-threat-assessment-iocta-2016
Europol: The Internet Organised Crime Threat Assessment (IOCTA) (2018). https://www.europol.europa.eu/internet-organised-crime-threat-assessment-2018
Europol: The Internet Organised Crime Threat Assessment (IOCTA) (2019). https://www.europol.europa.eu/activities-services/main-reports/internet-organised-crime-threat-assessment-iocta-2019
Felson, M.: Crime and Everyday Life, 3rd edn. Thousand Oaks, CA, Sage (2002)
Felson, M., Clarke, R.V.: Opportunity Makes the Thief: Practical Theory for Crime Prevention. Police Research Series, Paper **98**. Home Office, London (1998)
Finnie, T., Petee, T., Jarvis, J.: The future challenges of cyber-crime. In: Proceedings of the Futures Working Group, vol. 5. Quantico, Virginia (2010)
Flaxman, S., Goel, S., Rao, J.: Filter bubbles, echo chambers, and online news consumption, Public Opin. Q. **80**, 298–320 (2016) (Special Issue, 2016)
Fogg, B.J.: Persuasive Technology: Using Computers to Change What We Think and Do. Morgan Kaufmann (Dec 2002)
Gordon S., Ford, R.: Cyberterrorism? In: Cybterterrorism. The International Library of Essays in Terrorism, Alan O'Day, Ashgate (2004)
Gordon S., Ford, R.: On the definition and classification of cybercrime. J. Comput. Virol. **2**(1), 13–20 (2006)
Hargreaves, C., Prince, D.: Understanding cyber criminals and measuring their future activity. Lancaster University, Lancaster (2013)
Hutchings, A., Hayes, H.: Routine activity theory and who gets caught in the 'Net'? Curr. Issues Crim. Justice **20**(3), 433–445 (2009)
International Telecommunication Union (ITU): Recommendation ITU-T X.1205. Overview of cybersecurity, (April 2008). https://www.itu.int/rec/T-REC-X.1205-200804-I
Kruglanski, A.W., Thomson, E.P.: Persuasion by a single route: A view from the unimodal. Psychol. Inq. **10**, 83–109 (1999)
Lastdrager, E.E.H.: Achieving a consensual definition of phishing based on a systematic review of the literature. Crime Sci. **3**, 1–6 (2014)
Leukfeldt, E.R. (ed.): Research Agenda the Human Factor in Cybercrime and Cybersecurity. Eleven International Publishers, The Hague (2017)
Madarie, R.: Hackers' motivations: testing Schwartz's Theory of motivational types of values in a sample of hackers. Int. J. Cyber Criminol. **11**(1), 78–97 (2017)
Marwick A., Lewis R.: Media Manipulation and Disinformation Online. Data & Society Research Institute (2017)
Martens, B., Aguiar, L., Gomez-Herrera, E., Mueller-Langer, F.: The digital transformation of news media and the rise of disinformation and fake news-an economic perspective; Digital Economy Working Paper 2018-02; JRC Technical Reports (2018)
Martin, K.: Ethical implications and accountability of algorithms. J. Bus. Ethics 160, 835–850 (2019)
McGuffin, C., Mitchell, P.: On domains: cyber and the practice of warfare. Int. J. **69**(3), 394–412 (2014)
Morozov, E.: To Save Everything, Click Here. The Folly of Technological Solutionism, Public Affairs (2013)

McCarthy, J., Minsky, M.L., Rochester, N., Shannon, C.E.: A proposal for the Dartmouth Summer Research Project on Artificial Intelligence' (1955). http://www-formal.stanford.edu/jmc/history/dartmouth/dartmouth.html
McKinsey Global Institute: Artificial Intelligence, the Next Digital Frontier? Discussion Paper (June 2017)
Nielsen, J.: IM, not IP (Information Pollution). ACM Queue **1**(8), 75–76 (2003)
NIST: Glossary of Key Information Security Terms, Kissel, R. (ed.) (May 2013)
NATO: Cyber defence concept MC0571. Brussels, Belgium (2008)
NATO Strategic Communications Centre of Excellence: Social media as a tool of hybrid warfare, Riga (2016). https://www.stratcomcoe.org/social-media-tool-hybrid-warfare
Patchin, J.W., Hinduja, S.: Bullies move beyond the schoolyard: a preliminary look at cyberbullying. Youth Violence Juv. Justice **4**, 148–169 (2006)
PGG 2018: Cybercrime@EAP Project, Cybercrime and cybersecurity strategies in the Eastern Partnership region. Results of series of national workshops Eastern Partnership countries (2019)
Pladna, B.: Cyber terrorism and information security (2008). Accessed 12 Dec 2012. http://www.infosecwriters.com/text_resources/pdf/BPladna_Cyber_Terrorism.pdf
Pratt, T.C., Holtfreter, K., Reisig, M.D.: Routine online activity and internet fraud targeting: extending the generality of routine activity theory. J. Res. Crime Delinq. **47**(3), 267–296 (2010) (Wilson, C: Computer Attack and Cyber Terrorism: Vulnerabilities and Policy Issues for Congress. CRS Web, 2003)
Reid, R., van Niekerk, J.: From information security to cyber security cultures organizations to societies. Inf. Secur. S. Afr. (ISSA) **2014**, 1–7 (2014)
Report Flashpoint: Pricing Analysis of Goods in Cybercrime Communities (2019). https://go.flashpoint-intel.com/docs/pricing-analysis-of-goods-in-cybercrime-communities
Seebruck, R.: A typology of hackers: classifying cyber malfeasance using a weighted arc circumplex model. Digit. Investig. **14**, 36–45 (2015)
SELMA: Hacking Online Hate: Building an Evidence Base for Educators (2019). www.hackinghate.eu
Shariff, S., Hoff, D.L.: Cyberbullying: clarifying legal boundaries for school supervision in cyberspace. Int. J. Cyber Criminol. **1**(1), 76–118 (2007)
Schatz, D., Bashroush, R., Wall, J.: Towards a more representative definition of cyber security. J. Digit. Forensics Secur. Law **12**(2), Article 8 (2017)
Shipley, T.G., Bowker, A.: Investigating Internet Crimes. An Introduction to Solving Crimes in Cyberspace, Syngress (2014)
Spitzberg, B.H., Hoobler, G.: Cyberstalking and the technologies of interpersonal terrorism. New Media Soc. **4**, 71–92 (2002)
Stalans, L.J., Finn, M.A.: Understanding how the internet facilitates crime and deviance. Vict. Offenders **11**(4), 501–508 (2016)
Susarla A.: How artificial intelligence can detect–and create–fake news, The conversation, 3 May 2018. https://theconversation.com/how-artificial-intelligence-can-detect-and-create-fake-news-95404
Turkle, S.: Alone together: why we expect more from technology and less from each other. Basic Books, New York, NY (2011)
UNODC: Comprehensive Study on Cybercrime, United Nations, New York (2013)
Verizon: Data Breach Investigations Report (2018). https://enterprise.verizon.com/resources/reports/DBIR_2018_Report.pdf
Von Solms, R., Van Niekerk, J.: From Information Security to Cybersecurity. Computer & Security, Elsevier Ltd., vol. 38, pp. 97–102 (2013)
Wamala, F.: ITU National Cybersecurity Strategy Guide (2011). http://www.itu.int/ITU-D/cyb/cybersecurity/docs/ITUNationalCybersecurityStrategyGuide.pdf
Wardle, C., Derekshan, H.: Information disorder: toward an interdisciplinary framework for research and policy making, Council of Europe report, DGI(2017)09 (2017)
WEF: World Economic Forum–Forum Global Risks Report (2019)

WEF (World Economic Forum): Personal Data: The Emergence of a New Asset Class (2011). http://www3.weforum.org/docs/WEF_ITTC_PersonalDataNewAsset_Report_2011.pdf
Wiener, N. Cybernetics: Or Control and Communication in the Animal and the Machine, MIT Press, Cambridge, MA (1948)
Willard, N.E.: Cyberbullying and Cyberthreats: Responding to the Challenge of Online Social Aggression, Threats, and Distress. Research Press, Champaign, Illinois (2007)
Zhen, J., Zhao, L., Yan, J.: Why would people purchase personalized products online? An Explorative study. J. Inf. Technol. Manag. **4**, 18–30 (2017)

# Chapter 2
# Security: Human Nature and Behaviour

**Abstract** The human factor has always been considered the weakest link in the security chain. In this chapter we use a social psychology approach to discuss people's behaviour in relation to cybersecurity, by considering human errors, personality traits, the relationship between attitude and behaviour and the influence of social and situational factors. Human error has been widely studied in literature, especially in aviation and health care fields. Regardless of the area involved, analysing human factors is fundamental to understand the causes of accidents. With respect to cybersecurity, in fact, human errors—deriving from, e.g., work pressure, distraction, lack of awareness, organizational factors—can be considered one of the most important causes of security breaches. Moreover, looking at the relationship between attitude and behaviour, we can see how individuals do not always behave in coherence with their beliefs. For instance, regarding the protection of personal data, while many people are convinced that this is absolutely needed, in practice they continue to share data and information on social media, underestimating the related risks and consequences. A global perspective in investigating human factors within organizations requires considering the work environment. Furthermore, an increasingly technological context exposes individuals to different stimuli and stressors, generating critical conditions—such as information overload and technostress—affecting individuals' behaviour. Therefore, for effective adoption of secure behaviour, the management of employees' well-being has to be included in cybersecurity awareness programmes.

## 2.1 Human Error

Literature on human error is extensive, especially in sectors such as health care and aviation; in fact, in these complex environments doctors, pilots, and teams in general, interact with technology and have to deal with a variety of sources of threats (Helmreich 2000). In both areas important studies and models have been developed to prevent the dramatic consequences determined by human error.

© Springer Nature Switzerland AG 2020

I. Corradini, *Building a Cybersecurity Culture in Organizations*,
Studies in Systems, Decision and Control 284,
https://doi.org/10.1007/978-3-030-43999-6_2

Dealing with human error is a significant part of "human factors" research, whose development involves different areas which features many definitions. For instance, according to the International Ergonomics Association (IEA)[1] human factors is the:

> scientific discipline concerned with the understanding of interactions among humans and other elements of a system, and the profession that applies theory, principles, data and methods to design in order to optimize human well-being and overall system performance

Understanding how humans and systems interact helps to improve efficiency, productivity, job satisfaction and quality of life, and to reduce errors. There is, in fact, a general tendency to perceive human factors negatively, pointing especially at errors and the relative consequences.

In Reason's well-known model (1990, 1997), the concept of error refers to the failure to complete a planned series of mental actions or to the wrong planning in achieving a desired goal that cannot be attributed to chance. Classification on human errors includes slips, lapses and mistakes. Slips consist of failures in the execution of the intended action; lapses are memory failures; mistakes are decision-making failures, even if they are correctly executed. Intentions, subsequent actions and goals achieved are therefore important to classify an error as a slip or mistake (Norman 1981).

Despite most errors being unintentional, people can decide to do the wrong thing deliberately (intentional actions), such as in the case of violations, whose impacts on individuals and organizations can be extremely severe.

In health services a human-based approach is fundamental to reduce risks and prevent future errors (WHO 2016). In particular, medication-related error is one of the most common types of error (Kohn et al. 2000), and it is responsible for high costs. We cannot ignore the enormous losses generated by a single aviation accident; this is why human factors in aviation is a significant research area. The Human Factors Analysis and Classification System—HFACS (Shappell and Wiegmann 2001; Wiegmann and Shappell 2003) is one the most popular frameworks for the analysis of aviation accidents, and it is based on the Reason's model cited above. Accidents are the result of various causes; in fact, even though pilot's errors are often associated to mistakes in decision-making process (tactical errors) or to the lack of proper training (operational errors), many other factors contribute to their explanation. In particular, it is necessary to consider both individual factors, such as emotional and mental state, fatigue, cognitive bias, physical state, and the social environment, like adequate communication between crew members or crew negligence.

The combination of individual and environmental factors is the best way to explain workplace accidents, too. Individual stressors, inadequate communication about procedures, nature of the job, maturity safety cultures, and effective leadership are involved in the analysis of accidents at work. Moreover, work conditions can be "error-likely" (Swain and Guttman 1983), when they involve demands on human beings not compatible with their capabilities.

---

[1] https://www.hfes.org/resources/educational-and-professional-resources/new-item.

The cognitive effort is an important element to understand how errors work. In this sense, the Skill-Rules-Knowledge model (Rasmussen 1983) describes human mental approach to handle daily situations. Specifically, performance at skill-, rule-, and knowledge-based levels require different cognitive efforts.

At Skill level, the routine behaviour is based on our experience and learned skills, therefore the cognitive effort required is low. We can say that we act automatically, since we are fully familiar with the situation; for example, when we drive or walk towards a usual destination we often take a direction automatically, without subsequently remembering exactly which route was taken. The main errors associated to this situation are slips and lapses, supported by individual factors like fatigue and anxiety.

At the Rule level, behaviour is driven by rules that individuals use to perform known tasks. This condition is characterized by a higher cognitive effort than Skill level, and errors are often connected to the procedures chosen; whether partially correct or not, these procedures are not adequate for handling the specific situation.

Finally, at the Knowledge level, individuals have to face new or unexpected situations, relying on knowledge, skills and experiences acquired. Given that handling these situations effectively depends on individual's creativity and flexibility, the cognitive commitment required is very high. Errors typically involved at this level relate to adopting wrong adopted procedures.

The importance of human factors and the issue of error in cybersecurity field have substantially increased over the last few years. Since human beings and technologies are more and more interconnected, it is obvious that security cannot be only managed as a technological problem. People should instead be the major concern in cybersecurity (e.g. Schultz 2005; Schneier 2000; Safa et al. 2015), given that human errors are one of the most important causes of security breaches.[2]

Practically, it is possible to identify several frequent human errors and insecure behaviour determined by human mechanisms. For example, a common issue is related to individuals falling for phishing emails: despite many alarm campaigns, individuals continue to click on unknown links and to open unsafe attachments.

The success of phishing emails is often connected to emotional factors (e.g. fear and curiosity) in association with other conditions (e.g. high pressure, low risk perception, lack of awareness).

Familiarity of tasks can lead to errors, too. For example, people tend to click on links because they perceive this activity as familiar, no different from other situations they experience. A familiar situation, like receiving emails, is able to generate errors, as well as the automatic completion of phrases or email addresses, determined by technology. This automatic action can be responsible for sending emails to wrong recipients, together with distraction, pressure and lack of control by human beings.

Organizations plan specific activities (penetration testing) to test their technical and human vulnerabilities. However, besides penetration testing it would also be useful to invest in productive training based on emotional experiences (Chap. 7).

---

[2]According to analysis of data from the UK's Information Commissioner's Office (ICO) by the company CybSafe, 90 per cent of cyber data breaches in 2019 were caused by user error https://www.cybsafe.com/press-releases/human-error-to-blame-for-9-in-10-uk-cyber-data-breaches-in-2019/.

Some errors are related to implementing incorrect procedures, such in the case of password management. Considering the amount of information and passwords that have to be handled every day,[3] this activity can be extremely tiring. Due to the effect of heuristic processes, individuals tend to choose passwords easy to remember but also easy to crack,[4] thanks to specific software which can guess them by automatically exploring the different combinations.

Despite the problem being widely known, most people continue to handle passwords inappropriately, writing them down on post-it, or sharing them with others. In doing so, they underestimate the risks they (and their organization) are exposed to.

Other types of error include: losing paperwork, storing data in an insecure location, using company devices out of work inappropriately. Such a situation is mainly associated to the availability of Bring Your Own Device (BYOD) and to flexible work conditions (e.g. smart working). Since sensitive data about companies and customers are stored in work devices, it is evident that their misuse can be dangerous for businesses.[5]

Many errors in cybersecurity are due to a combination of individual and organizational factors (e.g. Bratus et al. 2008; Kraemer and Carayon 2007; Kraemer et al. 2009; Safa et al. 2016). Among them, we find lack of awareness, distraction and boredom, poor management support, inadequate budgets for security awareness, stressful work environments.

This is why the study of the human factor necessitates the intervention of different disciplines; consequently, appropriate training programmes should consider all the different aspects involved.

Finally, it is clear that human error can be strongly limited but not completely eliminated. In this sense, an interesting debate concerns the possibility of relying on automated systems to handle human factor.

Obviously, humans make mistakes, but even machines are not 100% reliable. In aviation field, for example, the growth of automated systems does not necessarily ensure a growth of security.[6] Looking at the recommendations provided by Federal Aviation Administration (FFA 2013), we notice that it includes the encouragement to improve the awareness of auto flight mode.

---

[3]According to a report published by Cybersecurity Ventures in cooperation with Thycotic, the number of passwords will probably surpass 300 billion by the year 2020, in particular for the effect of connected devices and Internet of Things. As passwords multiply, so do the risks https://thycotic.com/company/blog/2017/01/31/new-report-finds-300-billion-passwords-will-risk-2020/.

[4]Analysing the list of most common passwords, it is surprising that people still use easy password like 123456, password, qwerty, and so on. See, for example, the survey conducted by UK's National Cyber Security Centre (NCSC) in 2019 https://www.ncsc.gov.uk/news/most-hacked-passwords-revealed-as-uk-cyber-survey-exposes-gaps-in-online-security.

[5]See, for example, the 2018 User Risk Report, based on a survey conducted more than 6,000 working adults across the US, UK, France, Germany, Italy, and Australia about their personal security habits https://www.proofpoint.com/us/security-awareness/post/employees-take-risks-when-using-corporate-devices-personal-tasks. According to this report, 55% of workers at home allows to trusted friends and family to use devices provided to them by their employer.

[6]See, for example, the Ethiopian plane crash that happened in 2019 and caused the death of 157 people. According to the investigations, the incident was not due to a human error, but to a faulty software. In fact, the pilot tried several times to disable the automatic flight control, which instead pushed the aircraft in a few minutes after take-off.

A machine is able to do exactly what it was programmed for, but it would not be able to respond effectively in cases of unexpected events and anomalies. Even though a machine can perform a specific task or carry out a deep analysis, there will always be something unpredictable that cannot be anticipated.

With the increasingly complex human-machine interfaces, we should not underestimate the excess of confidence on the part of human beings in the infallibility of technologies (EU-OSHA 2017); this could lead to decision errors at work.

An approach where human beings and machines work together is the best way of preventing risks.

## 2.2 Personality Traits

Research on victimization risk factors provides useful information about victims' vulnerabilities, which are exploited by cybercriminals for their activities.

Studies in this area are worth exploring, considering that personal and environmental characteristics are important factors able to induce individuals to fall victim to cybercrime (Jansen et al. 2017).

For instance, research has shown how personal traits, like neuroticism and openness, can contribute to online vulnerability (e.g. Halevi et al. 2013a, b), and explain why certain people are more likely to answer phishing emails or post more information on social media. However, we must consider the importance of other psychological factors together with personality traits in order to explain human vulnerability to deception: situational and contextual factors can alter the effectiveness of phishing attempts (Goel et al. 2017).

Personality is a complex concept referring to "individual differences in characteristic patterns of thinking, feeling and behaving" (APA, American Psychological Association),[7] and several theories and models have been applied to its study and measurement.

Despite different approaches, there is a common agreement on considering five distinct dimensions of personality for its description: the five-factor model of personality (Big Five) is a multidimensional representation of personality structure (McCrae and John 1992; Costa and McCrae 1992).[8] This model is based on five dimensions: openness, conscientiousness, extroversion, agreeableness, neuroticism (this approach is often cited with the acronym OCEAN).

Even though a complete explanation of personality cannot be based on five dimensions, these are considered fundamental, since they include many terms related to personality description. In fact, each dimension encompasses several traits, as described in the following table (Akerman 2017; Halevi et al. 2017) (Table 2.1).

Each dimension has a positive and a negative side: openness vs closedness; conscientiousness vs impulsivity; extraversion vs introversion; agreeableness vs disagreeableness; neuroticism vs emotional stability.

---

[7] https://www.apa.org/topics/personality/.

[8] NEO Personality Inventory, for its revised version see (Costa and McCrae 1992; McCrae et al., 2005), is a widely used measure of the five-factor model.

**Table 2.1** The five dimensions of personality

| Dimensions | Description | Common traits (examples) |
|---|---|---|
| Openness | Opening to experience, willingness to try new things | Creativity, curiosity, imagination |
| Conscientiousness | Tendency to control impulses and respecting rules | Control, scrupulousness, reliability, persistence, perseverance |
| Extraversion | Interacting with others drawing energy | Social confidence, dynamism, assertiveness, friendliness |
| Agreeableness | Personal orientation to others | Altruism, sensitivity, helpfulness, trust, moderation |
| Neuroticism | Tendency to experiment negative feelings and emotions | Nervousness, pessimism, anxiety, lack of confidence |

The five dimensions of personality identify how individuals differ from each other and provide an important contribute to the explanation of their behaviour. For instance, since conscientiousness is linked to how we control our impulses, it is obvious that a conscientious person tends to be more responsible and dutiful, and therefore less impulsive. However, behaviour also depends on different situations that an individual has to manage; for instance, it is true that there are people who are more extraverted than others, but it is also true that they show different degrees of extraversion according to the particular situation they experience. The discussion regarding personality traits is complex, since the five factors of OCEAN model are connected with many other aspects of our lives. Here, we just underline their importance in affecting people behaviour; in this sense, they have to be considered in order to understand why people act in a certain way in relation to security.

More in detail, personality traits are associated to different security behaviour. The investigation of this correlation (Egelman and Peer 2015) shows how people with a high level of inquisitiveness tend to adopt better security practices, while impulsivity is negatively associated to protective behaviour.

Risky cybersecurity behaviour appears to be linked to the addictive use of the Internet (Hadlington 2017) and to impulsivity, since this leads people to act without thinking of the consequences of their behaviour.

Moreover, factors like risk-taking preference, decision-making style and gender seem to be significant predictors of a good security behaviour (Gratian et al. 2018).

Other studies reveal a relationship between personality traits and information seeking (Heinström 2003): secure people have a positive attitude towards searching information; in addition, they are more willing to accept new information and possible changes.

Considering the importance of personality traits related to security behaviour, additional research should be developed. Moreover, it will be helpful to deploy cross-cultural studies, given that some personality traits affect individual's cybersecurity behaviour across different cultures (Halevi et al. 2017).

For organizations, understanding human vulnerabilities and characteristics associated to insecure behaviour can be useful in order to design appropriate cybersecurity awareness programmes and prevention strategies. Indeed, according to studies involving organizations, we observe that negligence is often connected to successful cyber-attacks, to the point where negligent users, especially those involved in phishing attacks, represent the primary concern for companies.[9]

Despite external threats being numerous and provoking media attention, the insider threat is still a relevant problem to be tackled within organizations.

The concept of insider is a broad one, since it includes (Schultz 2002; Pfleeger 2008) not only employees, but also contractors, consultants, third-party business partners, and former insiders who still have access credentials. Considering different data breach scenarios, five types of insider threats can be identified (Verizon 2019a):

- The "Careless Worker" represented by employees or partners who develop a variety of dangerous actions, like misappropriating resources and mishandling data;
- The "Inside Agent", recruited by external actors, with the goal of stealing company information;
- The "Disgruntled Employee", interesting in harming his/her organization, for example damaging data or business activity;
- The "Malicious Insider", who uses the existing privileges to access information for personal gain;
- The "Reckless Third-Party", represented by business partners who compromise security through negligence, misuse, and so on.

Regardless of specific motivation, insiders can count on their intimate knowledge of the organizational information systems and operations; hence, they take advantage of the concrete possibility to easily access information during routine work, bypassing security protocols (Bishop et al. 2009; Chinchani et al. 2005; Probst et al. 2010).

Insider motivation is essentially related to the type of the illegal action (Kont et al. 2015). Apart from unintentional insider, it is possible to identify different motivations for different actions: revenge for IT sabotage; financial need for internal fraud; dissatisfaction or financial need for espionage; new job position or foreign government for insider theft of intellectual property.

Insiders are somehow part of the organization; this means that their actions are affected by organizational and environmental factors (Predd et al. 2008), for instance: how the company sets up and communicates policy and security rules to their employees and what security awareness initiatives management decides to implement. Therefore, in handling insider threats, it is necessary to develop a strong organizational and cybersecurity culture.

---

[9]See White Paper: The 2019 What Keeps You up at Night Report, by KnowBe4 https://www.knowbe4.com/hubfs/The2019WhatKeepsYouUpatNightReport.pdf. The report, to which over 350 organizations globally participated to, is focused on six areas of concern: attack types; security initiatives; compliance security; user-related issues; resource issues; executive-level concerns.

## 2.3 Perceiving Security Risks

The concept of risk has been defined in numerous ways and has a different meaning for different people; however, the common denominator among various definitions is the distinction between reality and possibility (Sjöberg et al. 2004).

Risk perception is an important issue to understand people's behaviour, since it is involved in several daily activities. Individuals can decide to handle or to avoid a certain risk because of their subjective judgement.

In processing information from the environment, people are affected by their own frame of reference, including personal beliefs, experiences, habits and interpersonal communication. On one hand, risk perception is an individual process; on the other hand, it involves bias and experience in a complex combination (Renn and Rohrmann 2000).

In literature it is possible to identify several dimensions of the risk affecting individual's perception (Fischhoff et al. 1978; Slovic et al. 1982). Among these we find, for example: the immediacy of the effect, the catastrophic potential, how the risk makes people feel dread, how new or familiar the risk is.

Research has shown that often there is not correspondence between the objective risk and the subjective estimate (Slovic 2000); in this sense, people can be worried about activities which are not dangerous, but underestimate those activities that, instead, can results in dramatic consequences on their lives.

Heuristic processes and biases play an important role to explain how individuals perceive and handle risks; in particular, heuristics describe how people process information. Generally speaking, people tend to make evaluations using mental shortcuts; by doing so, they considering information partially. This is why they are often designated as "cognitive miser" (Fiske and Taylor 1984) because of their tendency in using simple rules (heuristics) in judgment and decision making, rather than sophisticated and analytical reasoning. For example, according to the availability heuristic (Kahneman et al. 1982; Tversky and Kahneman 1973), people judge situations and take decisions recalling to their minds those events that are more frequent or probable than others. When the partial information is in conjunction with strong emotions, like fear, people are induced to overestimate the frequency of risk (Paek and Hove 2017).

Sometimes heuristics are useful, sometimes not. They must not be interpreted as a bad functioning of our mind; nevertheless, shortcuts do not always guarantee the right choice to solve problems or make a decision, because of the lack of complete information about the situation.

Systematic errors in judgement can produce biases. Owing to cognitive bias, individuals might misperceive the frequency of risks, like in the case of optimistic bias (Weinstein 1980), referring to the tendency to believe that certain risks do not regard them but others.

Since risk perception is responsible for protective and dangerous behaviour, we can imagine the relevance of this theme for safety and security areas. Indeed, feeling personally at risk is an important prerequisite for the adoption of health-related protecting behaviour (Renner et al. 2015).

In the same way, it is not difficult to imagine how security risks perception can influence people's behaviour positively or negatively. We can think, for example, of the perception of invulnerability characterizing optimistic biases, and the consequent underestimation of the risk. Many people are more likely to consider the action of clicking on unknown link dangerous for others but not for them, because they are convinced to have complete control of their environment.

Overconfidence can lead people to overestimate their forecasting abilities, and results in serious consequences (Glaser and Weber 2007; Moore and Healy 2008). Due to the illusion of having fully ability of controlling situations they experience, people ignore that other factors—not under their direct control—can affect their management. On the contrary, criminals are aware of human vulnerabilities and take advantage of them for the success of their operations.

Cognitive biases can explain the relationship between risk perception and secure behaviour. In the "affect heuristic", especially studied in consumer behaviour (e.g. King and Slovic 2014), decision-making is influenced by emotions (e.g. pleasure, fear). This happens in particular when resources are limited or time to decide is running short. So, when people feel good towards a certain situation, they are more likely to have a lower risk perception (Schneier 2008).

Individuals often think they are aware of risks because of the familiarity with them, but this does not ensure protective behaviour: just think of people who continue to fall for phishing emails.

We can say that people lack "sensors" to confidently move in the digital context (Corradini and Nardelli 2017). We are used to touching objects experiencing the sensation of physicality; in a way, this experience can also be reproduced digitally through virtual reality, since these applications are becoming more and more interactive and able to involve different human senses. Sometimes, applications require understanding complex scenario, generating security problems, such as in the case of complex touch screen surfaces.[10]

Shortage of physicality in the digital world gives rise to a different perception of the effects of situations, too. In particular, consequences of computer-mediated actions are perceived differently from the physical ones: in fact, when they are tangible and immediate, individuals tend to consider them more serious. A plan crash, for example, has instantaneous and visible consequences. Generally speaking, the more sensational the visually effects are, the more we tend to overestimate the gravity of the event. This is why most people tend to consider the frequency of plane crashes higher than car accidents.

[10]The investigation of fatal USS John S. McCain collision (between a destroyer and an oil tanker) in 2017 showed that lack of training and fatigue were responsible for the accident, associated to a complex touchscreen system of control. The Navy has decided to revert destroyers back to a physical throttle and traditional helm control system. https://news.usni.org/2019/08/09/navy-reverting-ddgs-back-to-physical-throttles-after-fleet-rejects-touchscreen-controls.

Risk factors for cybercrime and traditional crime differ from each other (Kranenbarg et al. 2017), since both types of crime are committed in different contexts. Because of the nature of cybercrime, its impact is generally perceived less serious compared to traditional crime. The comparison between a bank robbery and an Internet fraud helps us to understand the different mechanisms through which we evaluate the consequences.

In a bank robbery victim has an interaction with robber; even though the crime does not produce physical damages, it is perceived as a stressful experience.

In an Internet fraud, instead, victim does not have to come into physical contact with perpetrator and this is usually associated to a limited perception of the gravity of crime. Victims often become aware of the online fraud only at a later phase, when they control their banking accounts, or when they receive a notification from their bank. Only then, they experience that online fraud can also happen to them, and not only to others.

Even when computer-mediated, actions on the Internet are real, since they are expressions of people's behaviour and produce tangible effects, whether positive or negative. Unfortunately, the psychological distance between attackers and victims, generated by the use of digital technology, has determined a dichotomy in people's perception between virtual and real.

This psychological distance is bound to increase with the development of AI systems (e.g., Cummings 2004; Scharre 2018). In the case of remotely controlled weapons, the lack of physical contact with the victim can reduce the barrier to perform the action and remove the relative responsibility.

In psychology cognitive mechanisms of moral disengagement are well-known (Bandura 1999, 2016): they are moral standards which allow individuals to justify their unethical actions and preserve their self-image. In this sense, criminals can cognitively reconstruct their behaviour, reducing their negative emotions in transgressing. Bandura identifies eight moral disengagement mechanisms:

- moral justification (behaviour is considered acceptable);
- euphemistic language (using a certain language to make a damaging action respectable);
- advantageous comparison (damaging behaviours are considered less severe compared to others);
- diffusion of responsibility (harmful actions are authorised by authorities and this reduce personal responsibility);
- displacement of responsibility (sharing responsibility lower the burden);
- dehumanization of victim (depriving victims of their human characteristics);
- attribution of blame (harmful behaviour is considered as caused by the victim);
- distortion of consequences (the effects of the harmful action are altered in order to minimize personal responsibility).

These mechanisms can also be applied to the area of cybercrime, encouraged by the possibility of using anonymity and by the mechanism of dehumanization of victim. For instance, we can think of drone warfare, where drones are programmed to kill suspect terrorists autonomously, without the support of human guidance (Bandura 2017).

Finally, we should not forget the importance of language, since the frequent use of "cyber" to indicate everything related to technology, even when it is something else, can affect people's perception.

Considering the implications of risk perception and its relationship with secure behaviour, organizations should include the human factor in their cybersecurity strategy; not only training, but a real investigation of their employees' cyber-risks perception in their risk assessments (Corradini and Nardelli 2019).

## 2.4 Attitude Towards Security

Can a positive attitude towards security lead individuals to secure behaviour?

This is not a simple question, since the relationship between attitude and behaviour is rather complicated. This issue has been widely studied in literature, in particular in social psychology, the discipline focused on the interaction between people and their environment.

Attitude refers to individual tendency "to evaluate a particular entity with some degree of favour or disfavour" (Eagly and Chaiken 1998), hence it is internal; behaviour, instead, is what a person acts on materially, hence it is external. Attitude includes three components collectively named "the ABC model" (Rosenberg and Hovland 1960): affective (i.e., what we feel), behavioural (how we act), and cognitive (our beliefs and thoughts).

Attitude is often considered predictive of behaviour, but this is not an automatic process. Having a certain attitude does not guarantee its consistency with the relative behaviour. Firstly, people do not behave rationally at all times; moreover, personal and situational factors contribute to explain the processes through which attitude drives behaviour.

Important theories have tried to explain the relationship between them. The "theory of planned behaviour" (TBP, Ajzen 1985) is one of the most popular approaches to investigate factors affecting behavioural actions. This theoretical framework is based on the interaction of three elements (attitudes, subjective norms and the perceived behavioural control), which explain the reasoned decision-making process. This theory can be considered as an evolution of the theory of reasoned action (Ajzen and Fishbein 1980), which does not include the perceived behavioural control. However, behaviour cannot always be the result of a deliberate cognitive process, especially when actions are not significant and time is limited. In this situation, attitude can affect behaviour immediately, according to the MODE model (Fazio 1990).[11]

---

[11] According to the MODE model (Motivation and Opportunity as Determinants), there are two paths which connect attitude and behaviour: one path takes place according to a spontaneous processing method; another is voluntary and requires a considerable cognitive effort. Opportunities (available

Other models try to explain behaviour change. The Health Belief Model (HBM) was initially applied to understand people's health behaviour (Rosenstock 1974; Becker 1974), and suggests that change behaviour depends on six constructs (Jones et al. 2015): risk susceptibility, risk severity, benefits to action, barriers to actions, self-efficacy and cues to action. Therefore, according to this model, balance between costs and benefits leads to a preventive action. Messages reinforcing these elements play an important role in the persuasive process.

The theme of communication is significant to contribute to change behaviour. Typically, the aspects of communication are important in prevention activities (Chap. 5); however, in order to be effective, messages have to consider several aspects, first of all the needs of the target. In this sense, just providing information on security risks is not enough to be successful, because this activity is connected to several conditions, such as (Bada et al. 2015):

- individuals have to be able to understand security information;
- this information has to be compliant with their values and beliefs;
- they have to be motivated to change their behaviour.

In the context of cybersecurity, the theory of planned behaviour has been applied in several studies, in order to analyse the nature of employees' behaviours in relationship with organizational business strategies (Nobles 2018).

Encouraging attitude towards protective behaviour is an important goal, which can be achieved if employees believe that the expected consequences are positive; on the contrary, the conviction of a negative performance will produce an adverse attitude.

Therefore, the real challenge is to train people to strengthen the right attitude towards security, and not to add "security patches" when an incident occurs. We need to foster a cybersecurity mindset (Dutton 2017), which includes attitudes, beliefs and values, in order to motivate individuals to change their online behaviour.

Given the intricacy of human nature, it is evident that old behaviours do not change overnight (Hallas 2018). It takes time, and above all, we have to keep in mind that developing a long-term awareness requires absorbing the concept into the organizational culture (Corradini and Nardelli 2019) and considering all the obstacles and implications.

Moreover, we have evidence that relationship between attitude and behaviour is not so linear. An illuminating example is the use of social media regarding the protection of personal information.

According to some studies (Barth and de Jong 2017; Barnes 2006), while people claim to be very concerned about their privacy, their online behaviour does not align with what expressed. The so-called privacy paradox (Norberg et al. 2007) is a matter of utmost importance, since it reflects the disparity between the high value people give to their data and the lack of protecting behaviour, especially referring to the prudent use of social networks.

---

resources) and motivation to process information are the two fundamental factors to lead individuals to choose their path, and determine how attitude affects behaviour.

Despite the fact that risks related to the exposition of digital identity on the Internet and social media are well known, people have a natural tendency to share data and information. Several reasons explain this behaviour: relationship, love, friendly, emotional expression, sense of belonging to something.

Nevertheless, developing awareness is fundamental to increase the chance of careful actions, considering that people with a higher privacy awareness feel safer when using social networks sites (Bartsch and Dienlin 2016).

Such a situation affects not only younger people, but also adults. We can think, for example, of the “sharenting” phenomenon, which denotes parents’ behaviour in sharing online too many details about their children’s lives (Steinberg 2017). In this case, they overuse social media to share pictures and contents, infringing the privacy of children and putting them at risk, now and in the future (Brosch 2018).

Finally, the issue of privacy and people’s behaviour is particularly relevant in the application of wearable devices (Williams et al. 2019), whose diffusion is sharply increasing. Hence, people’s control on their own personal data is not so strong (Madden and Rainie 2015).[12]

One of the reasons explaining the careless behaviour on social media could be the information asymmetry between who gathers information and who shares it (Smerilli 2017). Even if individuals know that their data is used for profiling them, they are not aware of how this process works. They even help this activity unconsciously: because of biases in decision making process, when people are requested to fill documents with pre-set response options, they usually tend to confirm what they receive by default. The alternative, in fact, is to consciously read item by item and give or remove approval.

Besides the lack of awareness of the risks, we also have to consider that most people strongly desire to show that they are present on social media: posting information is equivalent to communicate to others that they exist and have something to say; in the same way, receiving likes on Facebook, Twitter or Instagram offers them an immediate gratification.

Some popular events which have compromised users’ data have determined a renewed awareness about the power of social media: the Facebook-Cambridge Analytica scandal in 2018 showed how about 87 million user’s personal data of Facebook profiles have been harvested and improperly used for political goal, without the consensus of people involved. At the same time, people do not seem to have changed the privacy settings they had before the event.[13]

[12]According to a research conducted by Pew Research Center (USA, 2015), 93% of American people say that the control over personal data is very important, but only 9% of them feel they have a lot of control over how much information is collected and how it is used. https://www.pewresearch.org/internet/2015/05/20/americans-views-about-data-collection-and-security/.

[13]See, for example, the survey conducted by The Atlantic; this is not a scientific study, but it is however interesting to understand how individuals perceive the importance of protecting their data in the digital world https://www.theatlantic.com/technology/archive/2018/06/did-cambridge-analytica-actually-change-facebook-users-behavior/562154/.

All in all, for many people their experience on the Internet can justify some risks. It is a fact that more and more commercial platforms offer suggestions to strengthen the visibility of individuals and companies on the Internet.

## 2.5 The Presence of Others

People's behaviour can be affected by the presence of other people or by their expectations. To understand the strength of this concept, we can refer to the definition of social psychology influence given by Allport (1954): "the attempt to understand and explain how the thought, feeling, and behaviour of individuals are influenced by the actual, imagined, or implied presence of other human beings".

Influence, therefore, depends not only on the physical presence of others: the pressure of social norms and expectations can, in fact, affect people's behaviour. Different forms of social influence have been studied in literature, like conformism (Asch 1955) and obedience (Milgram 1963). The difference between them is that while conformism refers to a change of attitude or behaviour in reaction to an implicit social norm, in the obedience the modification is the answer to an explicit order.

The social influence can also explain the adoption of positive or negative behaviour in work environment, when employees decide to comply with other colleagues' behaviour, regardless of company policies. We can think, for example, of the use of protective equipment: it is not rare to hear workers complaining that wearing these protective measures make work harder, or that working in a certain way has never produced negative consequences.

By doing so, they justify their behaviour; the critical issue is that their personal experience becomes a strong obstacle to the change of their behaviour. In particular, seniors typically show a strong resistance to new rules, because of the strength of their values and experiences, and this can affect juniors' actions; for instance, when seniors decide not to comply with safety rules, e.g. not using protecting equipment like helmets and goggles, they set a bad example, which could be imitated by juniors. In reality, since they have never had any accidents, and they have a lot of experience, why should they wear protective equipment?

We also have to consider that a junior's wish is to be accepted by the group of colleagues, and not isolated or discriminated. Being part of a group is thus more important than the practical application of what has to be done. Imitating others' behaviour and sharing the same experience are useful activities to achieve the goal.

The influence exerted on young workers is normative, because it is motivated by the wish to have approval and acceptance. This is different from the informational influence that, in the conformism phenomenon, describes how people assume positions expressed by others when they experience situations of uncertainty.

Some mediator factors can increase or reduce the effect of social influence, such as group cohesion, the dimension of the group, and social support. If these factors have a major effect on normative influence, other elements affect more informational influence, both individual (self-esteem) and situational (cultural norms).

On the other hand, organization is a social environment, where people look at what other people do and where the role of a positive leadership can make the difference. Hence, if managers do not pay attention to cybersecurity risks, it is likely that employees replicate the same behavioural pattern. People are more affected by what they see in practice around them rather than what they are told (Leach 2003).

Moreover, we have to consider the importance of norms and rules as part of work environment, since they both are social constructs: if they form the basis of our society, they are also part of working organizations. The difference is that rules are imposed by authority figures, while norms are not coercive, and they work if there is agreement on them. Norms are important because they identify the type of behaviour and mental pattern more or less acceptable by others, hence they determine the functioning of the organization.

When employees are encouraged to share norms, they can contribute to reinforce organizational culture. Not by chance, talking of Cybersecurity Culture, the starting point is the concept of organizational culture, which includes people's beliefs, values, attitudes and behaviour (Chap. 4). Hence, group norms and employees' behaviour can affect others (Roer 2015).

The value of these dynamics can be understood if we consider organizations as social structure, made up of beliefs and perceptions, and if we are convinced that the combination of organizational constructs and individual ones is essential in order to shape information security behaviour (Flores and Estedt 2012). Given their nature, a socio-technical point of view is needed for their investigation.

We should not forget that the acceptance of norms and rules promotes the construction of organizational trust climate, which represents a strategic element for developing Cybersecurity Culture.

The concepts discussed in this paragraph help us to understand how employees adapt to the corporate culture, for better or worse, suggesting useful paths for changing people's mindset towards security.

## 2.6 A Stressful Work Environment

In investigating factors affecting individual's behaviour, we cannot forget the role of the environment, and in this case, the influence of work environment.

This refers to organizational factors, such as structure, design and facilities, as well as the social environment. The quality of a workplace affects people's behaviour, since it may determine their motivation and, consequently, their performance (Leblebici 2012).

It is therefore evident that a stressful work environment can have negative effects on the entire organization, while proper management of stress can ensure prevention of accidents and promote protective behaviour in security.

In literature, in fact, it is well-known that stress can be responsible for several human errors and accidents in workplaces (Hilton and Whiteford 2010).

In the same way, we have to consider the importance of managing stress for security issues, because of the pressure imposed by organization to their employees about the need of information protection: they are required to be compliant with security policies and technical obligations (Lee et al. 2016a), alongside daily tasks, generating a good deal of stress. Besides the reduction of productivity, security-related stress can also be associated to information security violations by employees (D'Arcy et al. 2014).

The issue of the relationship between stress and security does not regard only employees, since even top managers appear to be more vulnerable to social engineering attacks in stressful conditions (Verizon 2019b). Moreover, we should not forget the continuous request of up-skilling, making cybersecurity professionals overworked and stressed out.[14]

Such a situation represents a vulnerability for companies: cybercriminals have a better chance of conducting successful attacks given that, differently from security professionals, they do not need to get a certificate of their skills before trying to hack a company server.

Managing stress therefore is a significant subject to be included in the discussion on the building of Cybersecurity Culture (Chap. 4). If people are stressed on the job, how can we expect them to comply with security policies and engage protecting behaviour?

Literature on stress is very extensive and includes different perspectives, since the concept of stress can be interpreted as (Cox et al. 2000):

- a stimulus (engineering approach)
- reactions of the organism to different stimuli (physiological approach)
- a process depending on the relation between individuals and their environment (psychological approach).

Considering that people are not passive vehicles for stimulus and response (Mark and Smith 2008), the third approach appears more respondent to the current concept of stress. Transactional theories (Lazarus and Folkman 1984; Cox and McKay 1981) are focused on the concept of stress as a process, and on the role of individuals in perceiving stress situations.

In recent years the topic of work-related stress has become subject of particular interest and has been extensively researched. Indeed, the European legislation on health and safety at work (Directive 89/391/EEC) has been a driving factor for this interest, since in accordance with this Directive employers have precise obligations, such as periodical risk assessments and providing necessary measures to protect workers.

Moreover, the European Framework Agreement (8th October 2004) underlines the need of preventing and managing stress in organizations. Stress is defined as

> a state, which is accompanied by physical, psychological or social complaints or dysfunctions and which results from individuals feeling unable to bridge a gap with the requirements or expectations placed on them.

---

[14]See the survey "High Alert: Tackling Cyber Security Overload in 2019" conducted by Symantec in collaboration with Goldsmith and University of London. The study has involved over 3,000 CISOs and senior cybersecurity decision makers across UK, France and Germany.

There are many theoretical frameworks and studies on sources of stress in workplace (e.g. Cooper et al. 1998, 2001; Edwards 2008; Karasek and Theorell 1990). Factors causing stress include—among others—individuals' exposition to pressure, job insecurity, the nature of job, inadequate work-life balance, lack of support from management, existing conflicts between staff. Useful frameworks help us to identify the main work design factors which may relate to stress-related health issues.[15]

Handling stress conditions is crucial for organizations, because of the important consequences. Individuals can cope with short-term exposure to pressure, which can even be considered positive, but they have difficulties in coping with prolonged exposure to intensive pressure. From an organizational point of view, stress has an impact on their effective functioning, since it can generate loss of productivity, turnover, sick leave, together with the development of a hostile climate among workers.

According to European Agency for Health and Safety (EU-OSHA)[16] work-related stress represents one of the most common health problems in the work environment; considering the current economic and social conditions stress is more likely to be a bigger issue in years to come.

Among the stress-related challenges we also have to consider the involvement of digital technologies, given that Information and Communication Technologies (ICT) are profoundly affecting work environment. Indeed, the complex set of interactions between humans and machines has generated new opportunities, but also risks.

### *2.6.1 Information Technologies, Technostress and Overload*

The current work environment is more and more characterized by the increasing dependence of managers on IT, and the introduction of updated versions of software and hardware to perform various tasks (Ragu-Nathan et al. 2008), while the pace of work is determined by the race for efficiency and profit.

ICT encourages flexible forms of work and fosters the virtualisation of the work environment, as in the case of virtual offices and smart working. Agile organizational approaches are made possible by digital technologies, since connectivity helps people to work remotely. In this way, spaces and costs can be optimized; moreover, if properly organised, this type of work can promote work-life balance.

On the one side, opportunities offered by the use of ICT are extremely positive; on the other side, risks related to people's health and safety cannot be ignored. In fact, ICT has changed the nature of work to the point where, for example, answering emails, and often at any time, is an integral part of routine work. This activity, together with the participation to online meetings and web conferences—which often

[15]See, for example, the HSE's Management Standard (HSE) which covers six key areas of work design: Demands, Control, Support, Relationships, Role, and Organisational change https://www.hse.gov.uk/stress/standards/index.htm.

[16]European Agency for Safety and Health at Work (EU-OSHA) is the European Union information agency for occupational safety and health https://osha.europa.eu/en/about-eu-osha.

replace the in-presence meetings, reduce the available period that workers devote to everything else.

The permanent accessibility determined by the use of technological devices has encouraged the development of OSH (occupational safety health) risks (OSHA 2013), such as:

- stress symptoms due to excessive working hours, workload and increasing complexity of tasks;
- information overload and difficulty of distinguishing significant and insignificant information;
- stress of continuous up-skilling;
- less face-to-face contact, given the tendency to replace humans with virtual contacts.

In this sense, with the growth of the use of digital technologies, terms like "technostress" and "information overload" are becoming more and more popular. Especially in recent years, technostress has been discussed in different ways, passing from an approach based on technology to an approach focused on the psycho-social interaction between people and digital technology. Initially, in fact, the term technostress was introduced (Brod 1984) in order to define a modern disease caused by the use of the new technologies (at that time referring only to computers).

Gradually, the meaning emphasized the negative impacts on human health generated directly or indirectly by technology (Weil and Rosen 1997); the current tendency is to focus on the experience between individual and technological environment. Two experiences of technostress are associated with ICT (Salanova et al. 2013): techno-strain (anxiety, mental fatigue, scepticism, inefficacy beliefs related to the use of ICT) and tecno-addiction (a compulsive use of them).

Further conceptualizations of technostress have identified five components (Tarafdar et al. 2007):

- Techno-overload refers to how ICT forces users to work faster and longer.
- Techno-invasion underlines the invasive effect of ICT, given that users potentially can be reached anytime, making employees feel constantly connected.
- Techno-complexity describes situations where the complexity associated with ICT makes users feel inadequate with respect to their skills, and therefore, they are forced to spend time and effort in further learning.
- Techno-insecurity is associated to users' feelings about possible job loss, since new technologies are replacing them.
- Techno-uncertainty refers to contexts where continuous ICT changes and upgrades generate uncertainty, inducing employees to constantly educate themselves.

It is clear that human's cognitive adaptation to the swift change of digital technology is a delicate issue that cannot be solved easily.

Modern organizations require people to adapt to the pace of technology, and not the contrary. So, workers have to answer huge volumes of emails and update their knowledge and skills to perform their tasks. Clearly, given the bounds on cognitive capacity of human beings, the larger the amount of information is, the less individuals are able to focus their attention on relevant ones.

According to research, continuous interruptions in workplace can have different impacts, being related to several factors, such as timing, the extent of interruptions, and individual's differences in their management (Mark et al. 2008): people try to compensate for interruptions by working faster, and by doing so, they experience more stress. We have to admit that interruptions produced by emails have a negative impact on employees' concentration, and therefore they represent a source of stress (Barley et al. 2011).

Research also considers the overwhelming volume of social demands deriving from social networking services (Lee et al. 2016b), since this can cause fatigue and psychological strain.

Overload associated to the use technology is a complex concept, which includes several components and exposes individuals to several health risks. Considering key elements of technology overload (Karr-Wisniewski and Lu 2010), we can distinguish between:

- information overload (too much information to process),
- communication overload (too many sources of communications to manage),
- system feature overload (the complexity of use).

Handling these types of overload is fundamental, since they can affect productivity and decision-making processes, individually and collectively. Moreover, considering that too much information can create confusion, employees can feel unsecure of where to focus their attention. The main possible consequence is that cues and signals of risk are ignored.

In conclusion, even though future research is necessary to deepen the relationship between ICT and stress (Berg-Beckhoff et al. 2017), we think that prevention of technostress and information overload is a fundamental issue for companies, also in relation to cybersecurity problems. Organizations should renew their stress risk assessment keeping in mind the increasing pervasiveness of ICT. Moreover, they should encourage mindful usage of technology in different ways (Colbert et al. 2016), for example experimenting technology-free meetings and face-to-face communication, reducing the use of emails. In this way, technologies can be perceived less pervasive and more collaborative.

We are aware that it is difficult to step back, but the relationship between people and technology has to achieve a positive balance in order to produce beneficial effects both for individuals and organizations.

### 2.6.2 Technology and Usability

Technological devices have changed how people work and interact with others. Nowadays, it is very difficult to separate personal and professional life, since digital instruments represent an extension of people's body and mind. The issue is that the more digitalized the world is, the more dependent on digital technology we are. We should not forget, for example, that digital technologies are designed to be persuasive and convince us to rely on them in order to make our lives easier and more comfortable. Regardless of the benefits produced by the current technologies, as humans, we need to rebuild a positive balance with them.

Notwithstanding more than 30 years of research in human-computer interaction (HCI) area, the concept that technological devices have to serve the needs of human beings (and not vice versa) is not yet taken for granted.

In this sense, the concept of usability is studied from different perspectives, such as computer science and psychology. Usability engineering focuses on improving the usability of interaction systems, providing a set of methods to guide and implement this kind of interaction (Nielsen 1993). This aspect has also to be considered in cybersecurity, since usability of security systems affects employees' security behaviour (Blythe 2013): poorly designed systems can induce people to bypass them and adopt insecure practices. In the same way, the design of interfaces can affect users' behaviour in installing security-related software updates (Mathur et al. 2016). Effective design can be achieved incorporating security activities into the natural workflow of daily tasks (Zinatullin 2016), through the involvement of IT professionals early in the design process.

When the term usability engineering began to develop around the mid of eighties, most people did not spend all their working hours using a PC and there was a clear separation between time spent interacting with IT devices and other activities. Now that our job is necessarily based on the use of digital devices, research is aptly speaking of "sustainable interaction", emphasizing the need of allowing human beings a proper management of their cognitive resources (Blevis 2007).

Tristan Harris, a designer who worked at Google, is convinced of the importance of recovering human value, in order to allow people to spend their time positively and to avoid the excess of distraction caused by the use of digital technologies. He explains how smartphones and social platforms use mechanisms similar to the functioning of slot machines: for both there is the risk of creating dependency in users. It is therefore fundamental to increase users' awareness of the "hijacking attention" determined by technologies.[17]

Considering the pervasiveness of digital technologies, it is not easy to find the right balance between individuals and them. However, we should start recognizing the responsibility of designers and producers, because we cannot afford that individuals are reduced to nothing more than cogs in the machine of technology.

Finally, we wish the adoption of more integrated approaches to ICT, encompassing separated psycho-social and technical viewpoints (Corradini and Nardelli 2014): this

---

[17] Harris created "The Center for Human Technology". In cooperation with Common Sense, a non-profit organization that deals with educating the appropriate use of digital technologies, an information campaign "Truth about tech" was launched in 2018 with the goal of reaching 55,000 public schools in the United States. Through the involvement of political decision makers, industries and public health experts, the intention is to push technological giants to rethink and redesign digital tools, considering their excessive intrusiveness and the risks associated with addiction. https://www.dailymail.co.uk/sciencetech/article-5352867/Campaign-launched-warn-dangers-technology.html.

means, above all, working on technical design and implementation of IT systems with an approach able to make them cognitively sustainable for individuals.

For this reason, cybersecurity cannot be considered just a technological issue, but it needs to pay great attention to the human factor.

## References


Ajzen, I.: From intentions to action: a theory of planned behavior. In: Kuhl, J., Beckman, J. (eds.) Action Control: From Cognitions to Behaviors, pp. 11–39. Springer, New York (1985)

Ajzen, I., Fishbein, M.: Understanding Attitudes and Predicting Social Behavior, Englewood CliVs. Prentice-Hall, NJ (1980)

Akerman, E.C.: Big Five Personality Traits: The OCEAN model Explained (2017). https://positivepsychology.com/big-five-personality-theory/

Allport, G.W.: The Nature of Prejudice. Addison-Wesley, Cambridge, MA (1954)

Asch, S.E.: Opinions and social pressure. Sci. Am. **193**, 31–35 (1955)

Bada, M., Sasse, M. A., Nurse, J.R.C.: Cyber security awareness campaigns: why do they fail to change behaviour? In: International Conference on Cyber Security for Sustainable Society (2015)

Bandura, A.: Moral disengagement in the perpetration of inhumanities. Personal. Soc. Psychol. Rev. **3**, 193–209 (1999)

Bandura, A.: Moral Disengagement: How People do Harm and Live with Themselves. Worth Publishers, New York (2016)

Bandura, A.: Disengaging morality from robotic war. Psychologist, February issue (2017)

Barley, S.R., Meyerson, D.E., Grodal, S.: Email as a source and symbol of stress. Organ. Sci. **22**, 887–906 (2011)

Barnes, S.B.: A Privacy Paradox: Social Networking in the United States. First Monday, vol. 11 (2006)

Barth, S., de Jong, M.D.T.: The privacy paradox: investigating discrepancies between expressed privacy concerns and actual online behaviour—a systematic literature review. Telemat. Inform. **34**(7), 1038–1058 (2017)

Bartsch, M., Dienlin, T.: Control your Facebook: an analysis of online privacy literacy. Comput. Hum. Behav. **56**, 147–154 (2016)

Becker, M.H.: The health belief model and personal health behavior. Health Educ. Monogr. **2**(4), 324–473 (1974)

Berg-Beckhoff, G., Nielsen, G., Larsen, E.L.: Use of information communication technology and stress, burnout, and mental health in older, middle-aged, and younger workers—results from a systematic review. Int. J. Occup. Environ. Health **23**, 160–171 (2017)

Bishop, M., Engle, S., Peisert, S., Whalen, S., Gates, C.: Case studies of an insider framework. In: Proceedings of the 42nd Hawaii International Conference on System Science, Jan 2009

Blevis, E.: In: Sustainable interaction design. In: International ACM Conference on Computer-Human Interaction (CHI), San Jose, CA, USA (2007)

Blythe, J.M. Cyber security in the workplace: Understanding and promoting behaviour change. In: Proceedings of CHItaly 2013 Doctoral Consortium, pp. 92–101 (2013)

Bratus, S., Masone, C., Smith, S.W.: Why do street-smart people do stupid things online? IEEE Secur. Priv. **6**(3), 71–74 (2008)

Breckler, S.J: Empirical validation of affect, behavior, and cognition as distinct components of attitude. J. Personal. Soc. Psychol. **47**(6), 1191–1205 (1984)

Brod, C.: Technostress: The Human Cost of the Computer Revolution. Addison Wesley, Reading MA (1984)

Brosch, A.: Sharenting—why do parents violate their children's privacy? New Educ. Rev. **54**(4), 75–85 (2018)

Chinchani, R., Iyer, A., Ngo, H., Upadhyaya, Q.S.: Towards a theory of insider threat assessment. In: The 2005 International Conference Dependable Systems and Networks (DSN'05), Yokohama, Japan, pp. 108–117. IEEE Computer Society, Washington, D.C. (2005)

Colbert, A., Yee, N., George, G.: The digital workforce and the workplace of the future. Acad. Manag. J. **59**(3), 731–739 (2016). Research Collection Lee Kong Chian School of Business

Cooper, C.L., Sloan, S.J., Williams, S.: Occupational Stress Indicator Management Guide. NFER-Nelson, Windsor (1998)

Cooper, C.L., Dewe, P.J., O'Driscoll, M.P.: Organizational Stress: A Review and Critique of Theory, Research and Applications. Sage Publications, CA (2001)

Corradini I., Nardelli, E.: Technostress prevention in digital society: for a new ecology of interaction between people and IT systems. In: International Conference on Digital Intelligence (DI-2014), Nantes, France, Sept 2014

Corradini, I., Nardelli E.: Sicurezza informatica: non vediamo e non vogliamo accettarlo. Link & Think Research Lab (2017). http://link-and-think.blogspot.com/2017/01/sicurezza-informatica-non-vediamo-e-non-vogliamo-accettarlo.html

Corradini, I., Nardelli, E.: Building organizational risk culture in cyber security: the role of human factors. In: Ahram, T., Nicholson, D. (eds.) Advances in Human Factors in Cybersecurity, AHFE 2018. Advances in Intelligent Systems and Computing, vol. 782. Springer, Cham (2019)

Costa, P., McCrae, R.R.: NEO PI-R professional manual Psychological Assessment Resources Inc. Odessa, FL (1992)

McCrae, Robert R., Costa, Jr., Paul T., Martin, Thomas A.: The NEO–PI–3: A more readable revised neo personality inventory, journal of personality assessment. J. Pers. Assessment **84**(3) 261–270 (2005)

Cox, T., Mackay, C.J.: A transactional approach to occupational stress. In: Corlett, E.N., Richardson, J. (eds.) Stress, Work Design and Productivity. Wiley, Chichester (1981)

Cox, T., Griffiths, A., Rial-Gonzalez, E.: Research on Work-Related Stress. Office for Official Publications of the European Communities, Luxembourg (2000)

Cummings M. L.: Creating moral buffers in weapon control interface design. IEEE Technol Soc Mag. **23**(3) (2004)

Cybersecurity Ventures: The World Will Need to Protect 300 Billion Passwords by (2020)

D'Arcy, J., Herath, T., Shoss, M.K.: Understanding employee responses to stressful information security requirements: a coping perspective. J. Manag. Inf. Syst. **31**(2), 285–318 (2014)

Dutton, W.H.: Fostering a cyber security mindset. Internet Policy Rev. **6**, 1 (2017)

Edwards, J.R.: Person-environment fit in organizations: an assessment of theoretical progress. Acad. Manag. Ann. **2**, 167–230 (2008)

Eagly, A.H., Chaiken, S.: Attitude structure and function. In: Gilbert, D.T., Susan T.F., Lindzey, G. (eds.) Handbook of Social Psychology, pp. 269–322. McGraw-Hill, New York (1998)

Egelman, S., Peer, E.: Scaling the security wall: developing a security behavior intentions scale (SeBIS). In: Proceedings of the ACM CHI'15 Conference on Human Factors in Computing Systems, vol. 1, pp. 2873–2882 (2015)

EU-OSHA: Key Trends and Drivers of Change in Information and Communication Technologies and Work Location (2017). https://osha.europa.eu/en/publications/key-trends-and-drivers-change-information-and-communication-technologies-and-work/view

Fazio, R.H.: Multiple processes by which attitudes guide behavior: The MODE model as an integrative framework. Adv. Exp. Soc. Psychol. **23**, 75–109 (1990)

Federal Aviation Administration (FAA): Fact Sheet—Report on the Operational Use of Flight Path Management Systems, Nov 21 2013

Fischhoff, B., Slovic, P., Lichtenstein, S., Read, S., Combs, B.: How safe is safe enough? A psychometric study of attitudes towards technological risks and benefits. Policy Sci. **9**, 127–152 (1978)

Fiske, S.T., Taylor, S.E.: Social Cognition. Addison-Wesley Publishing Company, Reading, MA (1984)

Flores, W. R., Ekstedt, M.: A model for investigation organizational impact on information security behavior. In: Seventh Annual Workshop on Information Security and Privacy (WISP) (2012)
Glaser, M., Weber, M.: Overconfidence and trading volume. Geneva Risk Insur. Rev. **32**, 1–36 (2007)
Goel, S., Williams, K., Dincelli, E.: Got phished? Internet security and human vulnerability. J. Assoc. Inf. Syst. **18**(1), 22–44 (2017)
Gratian, M., Bandi, S., Cukier, M., Dykstra, J., Ginther, A.: Correlating human traits and cyber security behaviour intentions. Comput. Secur. **73**, 345–358 (2018)
Hadlington, L.: Human factors in cyber security; examining the link between Internet addiction, impulsivity, attitudes towards cyber security, and risky cyber security behaviours. Heliyon **3**(7), e00346 (2017)
Halevi, T., Lewis, J., Memon, N.P.: Personality Traits and Facebook (2013a). arXiv:1301.7643v2
Halevi, T., Lewis, J., Memon, N.: A Pilot Study of Cyber Security and Privacy related Behavior and Personality Traits, 13–17 May. WWW 2013 Companion, Rio de Janeiro (2013b)
Halevi, T., Memon, N., Lewis, J., Kumaraguru, P., Arora, S., Dagar, N., Aloul, F.: Cultural and psychological factors in cyber-security. J. Mob. Multimed. **13**(1&2), 043–056 (2017)
Hallas, B.: Re-thinking the Human Factor: A Philosophical Approach to Information Security Awareness Behaviour and Culture. Hallas Institute (2018)
Heinström, J.: Five personality dimensions and their influence on information behavior. Inf. Res. **9**(1) (2003)
Helmreich, R.L.: On error management: lessons from aviation. BMJ **320**, 781–785 (2000)
HSE (2007)
Jansen, J., Junger, M., Kort, J., Leukfeldt, R., Veenstra, S., van Wiselm, J., van der Zee, S.: Victims. In: Leukfeldt, E.R. (ed.) Research Agenda. The Human Factor in Cybercrime and Cybersecurity. Eleven International Publisher, The Hague (2017)
Jones, C.L., Jensen, J.D., Scherr, C.L., Brown, N.R., Christy, K., Weaver, J.: The health belief model as an explanatory framework in communication research: exploring parallel, serial, and moderated mediation. Health Commun. **30**, 566–576 (2015)
Kahneman, D., Slovic, P., Tversky, A.: Judgment Under Uncertainty: Heuristics and Biases. Cambridge University Press, Cambridge, U.K. (1982)
Karasek, R.A., Theorell, T.: Healthy Work: Stress, Productivity, and the Reconstruction of Working Life. Basic Books, New York (1990)
Karr-Wisniewski, P., Lu, Y.: When more is too much: operationalizing technology overload and exploring its impact on knowledge worker productivity. Comput. Hum. Behav. **26**(5), 1061–1072 (2010)
King, J., Slovic, P.: The affect heuristic in early judgments of product innovations. J. Consum. Behav. **13**(6), 411–428 (2014)
Kohn, L., Corrigan, J., Donaldson, M. (eds.): To Err Is Human: Building a Safer Health System. National Academy Press, Washington, DC (2000)
KnowBe4, White Paper: The 2019 What Keeps You up at Night Report (2019). https://www.knowbe4.com/hubfs/The2019WhatKeepsYouUpatNightReport.pdf
Kont, M., Pihelgas, M., Wojtkowiak, J., Trinberg, L., Osula, A.M.: Insider Threat Detection Study. NATO CCD COE, Tallinn (2015)
Kraemer, S., Carayon, P.: Human errors and violations in computer and information security: the viewpoint of network administrators and security specialists. Appl. Ergon. **38**(2), 143–154 (2007)
Kraemer, S., Carayon, P., Clem, J.: Human and organizational factors in computer and information security: pathways to vulnerabilities. Comput. Secur. **28**(7), 509–520 (2009)
Kranenbarg, M.W., Holt, T.J., van Gelder, J.L.: Offending and Victimization in the Digital Age: Comparing Correlates of Cybercrime and Traditional Offending-Only, Victimization-Only and the Victimization-Offending Overlap, Deviant Behavior (2017)
Lazarus, R.S., Folkman, S.: Stress, Appraisal and Coping. Springer, New York (1984)
Leach, J.: Improving user security behaviour. Comput. Secur. **22**(8), 685–692 (2003)
Lee, C., Lee, C.C., Kim, S.: Understanding information security stress: focusing on the type of information security compliance activity. Comput. Secur. **59**, 60–70 (2016a)

Lee, A.R., Son, S.-M., Kim, K.K.: Information and communication technology overload and social networking service fatigue: a stress perspective. Comput. Hum. Behav. **55**, 51–61 (2016b)

Leblebici, D.: Impact of workplace quality on employee's productivity: a case study of a bank in Turkey. J. Bus. Econ. Financ. **1**(1), 38–49 (2012)

McCrae, R.R., John, O.P.: An introduction to the five-factor model and its applications. J. Pers. **60**(2), 175–215 (1992)

Madden, M., Rainie L.: Americans' Attitudes About Privacy Security and Surveillance. Pew Research Center (2015)

Mark, G.M., Smith, A.P.: Stress models: a review and suggested new direction. In: Houdmont, J., Leka, S. (eds.) Occupational Health Psychology: European Perspectives on Research, Education, and Practice, vol. 3, pp. 11–144. Nottingham University Press, Nottingham (2008)

Mark, G., Gudith, D., Klocke, U.: The cost of interrupted work: More speed and stress. In: Proceeding of the Twenty-Sixth Annual SIGCHI Conference on Human Factors in Computing Systems Association for Computing Machinery, 5–10 Apr 2008, Florence, Italy (2008)

Mathur, A., Engel, J., Sobti, S., Chang, V., Chetty, M.: They keep coming back like zombies: improving software updating interfaces. In: Symposium on Usable Privacy and Security (2016)

Milgram, S.: Behavioral study of obedience. J. Abnorm. Soc. Psychol. **67**, 371–378 (1963)

Mitnick, D.K., Simon, L.W.: The Art of Deception. Wiley Publishing, Inc (2011)

Moore, D.O., Healy, P.-J.: The trouble with overconfidence. Psychol. Rev. **115**, 2, 502–517 (2008)

Nielsen, J.: Usability Engineering. Morgan Kaufmann (1993)

Nobles, C.: Botching human factors in cybersecurity in business organizations. Holistica **9**(3), 71–88 (2018)

Norberg, P., Horne, D., Horne, D.: The privacy paradox: personal information disclosure intentions versus behaviors. J. Consum. Aff. **41**(1), 100–126 (2007)

Norman, D.A.: Categorization of action slips. Psychol Rev. **88** (1), 1–15 (1981)

Paek, H.J., Hove, T.: Risk Perceptions and Risk Characteristics, Oxford Research Encyclopedia of Communication (2017)

Pfleeger, C.P.: Reflections on the insider threat. In: Stolfo, S.J., Bellovin, S.M., Keromytis, A.D., Hershkop, S., Smith, S.W., Sinclair, S. (eds.) Insider Attack and Cyber Security. Advances in Information Security, vol. 39. Springer, Boston, MA (2008)

Predd, J., Pfleeger, S.L., Hunker, J., Bulford, C.: Insiders behaving badly. IEEE Secur. Priv. **6**(4), 66–70 (2008)

Probst, C.W., Hunker, J., Gollmann, D., Bishop, M.: Aspects of Insider Threats Insider Threats in Cyber Security, pp. 1–15. Springer (2010)

Proof Point, User Risk Report 2018. https://www.proofpoint.com/us/security-awareness/post/employees-take-risks-when-using-corporate-devices-personal-tasks

Reason, J.: Human Error. Cambridge University Press, Cambridge (1990)

Reason, J.: Managing the Risks of Organizational Accidents. Ashgate, Aldershot, U.K (1997)

Renn, O., Rohrmann, B.: Cross-Cultural risk perception research: state and challenges. In: Renn, O., Rohrmann, B. (eds.) Cross-Cultural Risk Perception, pp. 221–233. A Survey of Empirical Studies, Dordrecht, Boston (2000)

Renner, B., Gamp, M., Schmälzle, R., Schupp, H.T.: Health risk perception. In: Wright, James D. (ed.) International Encyclopedia of the Social & Behavioral Sciences, vol. 10, 2nd edn, pp. 702–709. Elsevier, Oxford (2015)

Ragu-Nathan, T.S., Tarafdar, M., Ragu-Nathan, B.S., Tu, Q.: The consequences of technostress for end users in organizations: conceptual development and empirical validation. Inf. Syst. Res. **19**(4), 417–433 (2008)

Rasmussen, J.: Skills, rules, and knowledge; signals, sings, and symbols, and other distinctions in human performance models. In: IEEE Transactions on Systems, Man, and Cybernetics, SMC-13, vol. 3, pp. 257–266, May–June (1983)

Roer, K.: Build a Security Culture. IT Governance Publishing (2015)

Rosenberg, M.J., Hovland, C.I.: Cognitive, affective and behavioral components of attitudes. In: Rosenberg, M.J., Hovland, C.I. McGuire, W.J., Abelson, R.P., & Brehm, J.W. (eds.) Attitude Organization and Change: An Analysis of Consistency Among Attitude Components, pp. 1–14. Yale University Press, New Haven (1960)

Rosenstock, I.M.: Historical origins of the health belief model. Health Educ. Monogr. **2**, 328–335 (1974)

Safa, N.S., Solms, R.V., Futcher, L.: Human aspects of information security in organisations. Comput. Fraud Secur. (2), 15–18 (2016)

Safa, N.S., Sookhak, M., Von Solms, R., Furnell, S., Ghani, N.A., Herawan, T.: Information security conscious care behaviour formation in organizations. Comput. Secur. **53**, 65–78 (2015)

Salanova, M., Llorens, S., Cifre, E.: The dark side of technologies: technostress among users of information and communication technologies. Int. J. Psychol. **48**(3), 422–436 (2013)

Scharre, P.: Army of None. Autonomous Weapons and the Future of War. W.W. Norton & Company, NY (2018)

Schneier, B.: Secrets and Lies. Wiley, New York (2000)

Schneier, B: The psychology of security. In: Proceedings of the Cryptology in Africa 1st International Conference on Progress in Cryptology (2008), pp. 50–79. Springer (2008)

Schultz, E.: A framework for understanding and predicting insider attacks. Comput. Secur. 526–531 (2002)

Schultz, E.: The human factor in security. Comput. Secur. **24**, 425–426 (2005)

Shappell, S.A., Wiegmann, D.A.: Applying reason: the human factors analysis and classification system (HFACS). Hum. Factors Aerosp. Saf. **1**, 59–86 (2001)

Sjöberg, L., Moen, B.-E., Rundmo, T.: Explaining risk perception. An Evaluation of the Psychometric Paradigm in Risk Perception Research. Norwegian University of Science and Technology. Rotunde Publications no. 84 (2004)

Symantec, High Alert: Tackling Cyber Security Overload in 2019, Chapter 1: Perfect Storm (2019)

Slovic, P., Fischoff, B., Lichtenstein, S.: Facts versus fears: understanding perceived risk (1982). In: Kahneman, D., Slovic, P., Tversky, A. (eds.) Judgment Under Uncertainty: Heuristics and Biases, pp. 463–489. Cambridge University Press, Cambridge, UK (1982)

Slovic, P.: The Perception of Risk. Earthscan, London, Sterling, VA (2000)

Smerilli, A.: Privacy, reputazione e comportamenti umani nell'era dell'informazione: questioni etiche ed osservazioni empiriche (2017), in Corradini, I. *Internet delle cose. Dati, sicurezza e reputazione*, Milano, Franco Angeli (2017)

Steinberg, S.B.: Sharenting: children's privacy in the age of social media. Emory Law J. **839**(66), 839–884 (2017)

Swain, A.D., Guttman, H.E.: Handbook of Human Reliability Analysis with Emphasis on Nuclear Power Plant Applications. NUREG/CR-1278. US Nuclear Regulatory Commission, Washington, DC (1983)

Tarafdar, M., Tu, Q., Ragu-Nathan, B.S., Ragu-Nathan, T.S.: The impact of technostress on role stress and productivity. J. Manag. Inf. Syst. **24**(1), 301–328 (2007)

Tversky, A., Kahneman, D.: Availability: A heuristic for judging frequency and probability. Cogn. Psychol. **5**, 207–232 (1973)

Verizon: Insider Threat Report (2019a). https://enterprise.verizon.com/resources/executivebriefs/2019/insider-threat-report-executive-summary.pdf

Verizon: Data Breach Investigations Report (2019b). https://enterprise.verizon.com/resources/executivebriefs/2019-dbir-executive-brief.pdf

WHO, World Health Organization: Human Factors, Technical Series on Safer Primary Care (2016)

Weil, M., Rosen, L.: TechnoStress: Coping with Technology @Work @Home @Play. Wiley, New York (1997)

Weinstein, N.D.: Unrealistic optimism about future life events. J. Pers. Soc. Psychol. **39**, 806–820 (1980)

Wiegmann, D.A., Shappell, S.A.: A Human Error Approach to Aviation Accident Analysis: The Human Factors Analysis and Classification System. Ashgate, Aldershot, UK (2003)

Williams, M., Nurse, J.R.C., Creese, S.: Smartwatch games: Encouraging privacy-protective behavior in a longitudinal study. Comput. Hum. Behav. **99**, 38–54 (2019)

Zinatullin, L.: The Psychology of Information Security: Resolving Conflicts Between Security Compliance and Human Behaviour. IT Governance Publishing, Ely, Cambridgeshire (2016)

# Chapter 3
# Redefining the Approach to Cybersecurity

**Abstract** One of the most critical issues in cybersecurity is represented by social engineering attacks. These threats have been known for years, but it is very difficult to handle them effectively, because they are strictly related to human nature. Social engineering is not just a phishing email; indeed, it is possible to distinguish several forms of attack which combine different elements, from human to social to physical and technological. According to a psychological point of view, social engineering is a powerful means of gaining information exploiting individuals' weaknesses. Moreover, due to the mechanisms of persuasion, widely studied in literature, it is easy to imagine how complicated the management of this threat is. Appropriate training of employees, especially of key roles of the company, can be an effective antidote to social engineering. Given the current scenario and the future perspective in cybersecurity, it is clear that the approach used to manage cybersecurity requires a radical change. Currently, the preferred cybersecurity strategy is still based on technological solutions, without brilliant results, since cyberthreats keep growing. Many are convinced that Artificial Intelligence (AI) will be an opportunity for managing cybersecurity; whether true or not, it is however evident that AI has also the power to generate new threats and to strengthen the existing ones. Therefore, we should be very prudent when technologies are presented as a miracle solution for cybersecurity problems. The starting point is that technology has to be deployed under full human control. Then, critical thinking is needed to develop alternatives to improve the current approach to cybersecurity. In short, we need to develop a multidisciplinary vision of cybersecurity, involving other disciplines and assuming different perspectives.

## 3.1 Social Engineering: The Real Trojan Horse of Cybersecurity

In the era of social media and digital communication, it is strange to admit that communication can become a threat: the more we are connected, the more we have to face dangers. One of the most critical issues in cybersecurity, based on communication techniques, is social engineering.

© Springer Nature Switzerland AG 2020

I. Corradini, *Building a Cybersecurity Culture in Organizations*,
Studies in Systems, Decision and Control 284,
https://doi.org/10.1007/978-3-030-43999-6_3

This concept seems to have been used for the first time in politics, and then migrated into cybersecurity (Hatfield 2017).

From a psychological viewpoint and in the context of cybersecurity, social engineering can be defined as a tactic which, using a persuasive communication, aims at gaining people's confidence in order to lead them to disclose sensitive information or to do something dangerous, e.g. to click a malicious link or to open an infected file.

If in this context social engineering has a negative meaning, we cannot say the same when applied to other fields. For example, a vendor frequently uses social engineering techniques to convince a potential buyer to acquire his products or services. A vendor needs effective communication abilities to appeal to individuals' emotions and persuade them. Hence, he uses empathy to understand clients' feelings. Also, empathizing with others in different life and work situations is a fundamental communication skill (Chap. 5).

The main element which characterizes social engineering is its power to manipulate people's perception by using different approaches. When communication is face-to-face, facial expressions and gestures are powerful means of persuasion. When the medium is digital technology, as in the case of phishing email, social engineering aims at generating a believable situation in order to capture people's attention.

Previously, phishing emails were easily identifiable, because of grammar errors or of the strange language used. Now, they are becoming more and more accurate with respect to contents and style; it is not therefore easy to recognize them, especially for inexperienced people, but even expert users can be confused by visual deception attacks (Dhamija et al. 2006). Currently, phishing represents a core attack method for all cybercrime (Europol 2019) and requires human interaction to succeed.

However, social engineering is not just a phishing email; we can distinguish several forms of attacks which may combine different aspects (Salahdine and Kaabouch 2019; Krombholz et al. 2015), i.e. human, physical, social and technical. In the following we describe some of them.

- Phishing: This is the most popular application of social engineering. It is the fraudulent practice of sending emails, usually appearing to come from a well-known source (e.g. an important organization) to steal sensitive information, like passwords, credit card numbers, etc.
- Spear phishing: It is a form of phishing tailored to the target recipient (individual or groups). Attackers study the behaviour of their targets and collect information to make the attack believable, in order to increase the likelihood of its success. When victims receive a spear phishing email, they think that the communication comes from a trusted source. For example, Business Email Compromise (BEC) is a fraud where the attacker impersonates an organization executive to lead an authorized employee in that organization to perform a wire transfer to an account controlled by the same attacker.
- Pretexting: Social engineering needs to achieve trust to be successful. For this purpose, it uses an appropriate scenario fabricated in order to convince a targeted victim. Pretexting consists, for example, of impersonating someone else, e.g. a police officer, or an insurance investigator.

- Tailgating: In this physical form of social engineering, someone gains access to a building or to a restricted area without proper authentication, but exploiting a convenient situation, for example, following another person entering the property.
- Whaling: In this form of phishing the main characteristic is the type of target, represented by senior executives, representative people of government agencies, politicians, and celebrities. Given the relevance of the target (big fish), the value of information is particularly attractive to cybercriminals. Like spear phishing, the scam email is tailor made, and appears to come from a business partner.
- Vishing: This is a phone scam which combines phishing and voice. Vishing can be considered the telephone equivalent of phishing. Here, given that the fraudulent action is over the phone, empathy and the ability of handling conversation are needed for the success of the attack.

We underline the multiplicity of the methods used in social engineering attacks, as well as the different levels of sophistication. For example, reverse social engineering (e.g. Irani et al. 2011) points to the active role of the victim: the attacker does not start the contact with her, but the victim herself is tricked and led to initiate the relationship with the attacker.

Apparently, social engineering techniques seem to be carried out spontaneously, especially when associated to the sending of massive amount of emails; there is always someone who falls for a phishing email.

Several phases define a social engineering attack (e.g. Mouton et al. 2016; Segovia et al. 2017); typically, it includes:

- target identification and information gathering;
- relationship development to gain the trust of the selected victim;
- execution, in order to exploit the trust achieved;
- exit, to avoid leaving proof and, at the same time, maintaining a good relationship with targets for future activities.

Whether or not it is happening through physical or technical means, the focus of social engineering is social interaction and manipulation. Understanding the mechanisms of persuasion is fundamental to handle phishing threat, since people tend to ignore the critical warning messages (Gupta et al. 2017), thus contributing to the success of the attack.

## 3.2 Persuasion in Social Engineering

Digital technologies and social media offer many opportunities to interact socially. Hence, a social engineer can interact with targeted people through social media platforms, and collect information directly posted by Internet users.

In investigating the psychological aspects of social engineering, we can say that this technique exploits both social and cognitive vulnerabilities (Corradini and

Nardelli 2020): if the social relationship can be -by its nature- a risk, the cognitive element[1] is represented by manipulation of people's perception.

In this sense, two forms of interaction can be identified for the manipulation, depending on how explicit the interaction is. When a victim gets in touch with the attacker (direct interaction), such in the case of phishing, spear phishing and vishing, email or telephone are the means which connect the two actors. In the indirect way, instead, manipulation works without starting an explicit relationship with victims. Posting false information on a website or on a social media, for example, can attract certain individuals or groups particularly interested in the published information. Here, the number of potential targets can be very high, even if false information is posted with the goal of capturing the attention of specific groups.

We know the effects of spreading fake news on the Internet and the difficulty of restoring proper communication. We also know how manipulation of information can alter people's perceptions and generate collective inadequate behaviour (Sect. 1.5). False and negative information on the financial market, for instance, could cause people's hysterical reactions, disrupt the economic balance and affect the financial relations among countries.

Threats based on social engineering have been known for years, but they continue to have high chances of success, because they are strictly connected to human nature. In short "[...] we, as human beings, are all vulnerable to being deceived because people can misplace their trust if manipulated in certain ways" (Mitnick and Simon 2002).

Indeed, investigating victimization in the case of phishing, we observe that individuals can have an active role in the success of social engineering attacks, given that for them it is usual to receive emails and phone calls, while it is difficult to judge messages in full detail to find markers of fraud (Jansen and Leukfeldt 2015).[2] Moreover, stress, pressure, and other factors can facilitate the lack of accurate control and ensure the success of the attack.

From a psychological point of view, it is interesting to explore how effective psychological principles of persuasion (Cialdini 1984, 2000) are. In this sense, understanding their function and promoting awareness on this subject should be included in cybersecurity education programmes.

In Table 3.1 we describe some principles characterizing social engineering attacks and some specific points of attention translated into simple questions.

The principles listed above work effectively because they are based on heuristic processes easily available to human minds (Sect. 2.3). They can count on the peripherical route of persuasion, which points to incidental cues, rather than to the strength

[1]In literature, social engineering is considered a form of cognitive hacking, that is "as gaining access to, or breaking into, a computer information system for the purpose of modifying certain behaviours of a human user in a way that violates the integrity of the overall user-information system" (Cybenko et al. 2004).

[2]The study consisted of an analysis of 600 phishing and malware incidents involving a Dutch bank, and focused on the behaviour of the customers victims of fraud. Despite the limitations of the study (regarding one bank), the authors underline how the combination of phishing (social engineering) and malware (technical skills) is becoming a more common method to commit fraud.

**Table 3.1** Persuasion principles and points of attention

| Persuasion principles | Description | Points of attention |
|---|---|---|
| Reciprocity | According to a social norm, individuals tend to return a favour: if someone gives something to others, they feel obliged to repay that debt | How selfless is the helper? |
| Commitment and consistency | People need to appear consistent in their behaviour. Hence, they act in line with their words or agreements, so as not to be perceived unreliable | Am I sensitive to people's judgment? |
| Social proof | People tend to do what the others do, especially in uncertain conditions | Do people behave in a certain way just for conformity? |
| Authority | People feel an obligation to obey figures of authority, even if they don't agree with them | Is the source reliable? |
| Liking | People tend to comply with requests made by those they like, because of physical attractiveness, familiarity and similarity | Is it really empathy? |
| Scarcity | People tend to consider things more valuable if less available. For example, finding offers available for a "limited time only", sales are encouraged | Do I really need it? |

Persuasion principles (Cialdini 1984, 2000) and points of attention in social engineering (Corradini 2017)

of the contents.[3] The power of persuasion is well-described in literature, especially in social psychology (e.g. Petty and Cacioppo 1986; Kruglanski and Thomson 1999).

The main factors of persuasion include:

- the communicator (*who*),
- the message content (*what*),
- the channel of communication (*how*),
- the audience (*whom*).

[3] According to literature, it is possible to distinguish two different methods of persuasion: the central route is characterized by the strength of the messages or of the arguments; to be effective, the person receiving the message must have a high motivation to listen to. Peripheral route, instead, do not require thinking carefully. For example, when we are distracted, we cannot be concentrated on the contents of a message, but we are attracted by superficial cues (depending on the context).

However, in investigating the persuasive process, we have to consider other many factors, such as: the characteristic of the source and its reliability; logical or emotional contents of the message and what they inspire; motivation and need of cognition; the flood of influence generated by the channels of communication.

It is evident that when factors are combined in a digital context (*where*), the space of influence is larger and more powerful (Corradini 2017). Digital technology has changed the nature of persuasion (Perloff 2014), increasing complexity and blurring the lines between three different concepts, such as information, influence and entertainment.

People are not often aware of the reliability of sources, and—unless they do a particular job—they do not bother to verify them. It is a fact that nobody is immune, to the point where even security experts can become victim of social engineering.[4]

Organizations have to consider this threat seriously, and to review training and awareness programmes to fight it (Aldawood and Skinner 2019): despite of their efforts, social engineering is still a significant problem for companies. Indeed, social engineering activity can be finalized to different targets with the purpose of stealing useful information. Specific positions can be targeted for this goal, for example executive assistants to the CEO, general managers, drivers, receptionists, the cleaning staff: according to their different roles, they handle sensitive information which needs to be protected.

Appropriate training combined with social engineering penetration testing can be a strong antidote to this persuasive form of communication; moreover, the points of attention identified in Table 3.1 are useful questions to invite people to reflect on the situation they experience, in order to ponder the different circumstances they have to face.

## 3.3 What Happens with Artificial Intelligence and Internet of Things?

As discussed in Chap. 1, the growth of Internet of Things (IoT) and Artificial Intelligence (AI) applications are defining a new technological environment, which produces advantages but also further security risks. Currently, many countries are more and more interested in deploying AI systems, to the point where the estimated business for 2030 is impressive.[5]

---

[4]See, for example, how Kane Gable, a 15-year-old, using social engineering, gained access to the personal and work accounts of some of America's most powerful spy chiefs.

The teenager persuaded call handlers at an internet giant that he was John Brennan, the then director of the CIA, to gain access to his computers and an FBI helpdesk that he was Mark Giuliano, then the agency's Deputy Director, to re-gain access to an intelligence database https://www.telegraph.co.uk/news/2018/01/19/british-15-year-old-gained-access-intelligence-operations-afghanistan/.

[5]According to McKinsey Global Institute AI has the potential to deliver additional global economic activity of around $13 trillion by 2030 https://www.mckinsey.com/featured-insights/artificial-intelligence/notes-from-the-ai-frontier-modeling-the-impact-of-ai-on-the-world-economy#part1.

It is evident that there is a strong reliance on automation; the application of AI in cybersecurity is accelerating: to fight against cyber-attacks, at least a third of Chief Information Security Officers (CISOs) have decided to adopt artificial intelligence (CISCO 2018). In addition, firms are more and more convinced that they need to implement AI systems to identity critical threats (Capgemini 2019).

It is more than likely that AI will re-design our lives, and that it will have a significant impact in a lot of fields. It is not very clear if this would have a positive or negative effect. Media and experts say that the growth of AI systems is going to improve our lives, because of the opportunities offered.

We always recommend great prudence when technology is presented as a miracle for our problems, before research clarifies benefits and risks. On the other hand, technology can be considered positively when human beings have its complete control.

Among recommendations on AI (Del Ponte 2018), it is recognized the need of discussing its societal risks, and the need of creating an ethical framework to regulate it, starting from the principle that this technology has to be developed under human control. In this sense, the Ethics Guidelines for Trustworthy AI by European Commission (2019) underlines the necessity of a full adherence of AI to humans' ethical principles and values (Sect. 1.1).

AI systems are developing rapidly, and the risk that their growth overcomes people's capacity of handling them is more than real. According to the approach of Machine Learning (ML), machines can "learn" without someone having to program them. And they learn rapidly, focusing on any data we, as humans, handle for our activities; so, if we search a certain product on different websites, we have to be aware that in the future similar products will be suggested to us.

It is obvious that humans cannot enter into competition with AI. The battle would be lost from the start. On the contrary, humans and machines have to work together respecting their own limits, because more and more in the future digital technologies will play an important role in our lives.

Hence, we have to exploit opportunities offered by the landscape of AI and, at the same time, we cannot ignore that AI increases relevant risks even in the workplace (Houghton and Green 2018), regarding health and safety, employee ethics, diversity and equality. One of the major risks is about discrimination: machine learning works on data, and there is no guarantee that this data is free from prejudice. Consequently, the algorithmic process of learning can build altered representations of reality on whose basis discriminatory answers are provided by the AI system.

Such an impact is significant when hiring people or promoting specific categories or genders for specific jobs; bias in predictive hiring tools is a relevant problem to be handled, since employers are more and more interested to use these predictive tools to reduce time and costs (Bogen and Rieke 2018).

In the meantime, precautionary measures are required when AI applications are proposed as the panacea for cybersecurity. They say that solutions based on AI will provide important support for the protection of organizations, ensuring effective security standards. AI will probably be able to identify new malware and cyberthreats, given its capacity to handle massive volume of data.

There are at least two considerations we should reflect on.

The first is that we are facing many threats not yet solved, despite using the most advanced technological solutions available on the market. Moreover, cyberthreats continue to evolve and improve, so every solution based on artificial intelligence needs to be constantly updated. Therefore, it is not clear why and how AI will be able to solve security problems without causing new ones.

The second issue is that, as with any innovation, everyone takes advantage. This means that criminals can also improve their "modus operandi" exploiting the scalable use of AI systems. We can assume an extension of the landscape of threats consisting of (Brundage et al. 2018)[6] a reinforcement of the existing menaces (e.g. expanding set of actors) and of the possibility of generating new ones.

From this point of view, we can say that social engineering techniques can be strengthened from AI systems. Indeed, since AI is able to mimic human voices realistically, a social engineer can create automated social engineering attacks, using recorded data and impersonating perfectly, even using the same style of language. No wonder that Facebook engineers have created a machine learning system named "MelNet" cloning the voice of Bill Gates,[7] and that a Canadian start-up introduced an AI system capable of synthesizing a person's voice from just a one-minute audio sample.[8]

In addition, the combination between AI and Internet of Things provides a powerful weapon for criminals. IoT devices make available a large amount of data, because their sensors are able to collect information about their environment. AI algorithms, on the other hand, can infer selected information from this data. Consequently, it is possible to produce accurate profiles of targeted people, as well as to identify further vulnerabilities.

Such a combination can also be useful for the implementation of social engineering attacks, where the ability of gathering information and profiling is essential for their success.

We are convinced that human beings are essential in cybersecurity, regardless of the wonderful technological solutions used. Who thinks that the employment of AI will completely replace human beings in the activity of protection has not yet understood the nature of the issue. Once again, the tendency is to underestimate the importance of human factors and of well-trained people, and to rely on AI completely. This conviction is the biggest threat to overcome.

Even if AI is able to recognize threats more quickly than humans, thanks to its ability to analyse a large amount of data, originality and human experience are quite difficult to replicate. This is especially true when it comes to tackling cyber-attacks:

---

[6]In this report the authors identify three representative domains: Besides digital security, and the problems of cyberattacks, they also consider the domain of physical security (e.g. the deployment of autonomous weapon systems) and the political security (the use of AI for propaganda and deception).

[7]https://www.theverge.com/2019/6/10/18659897/ai-voice-clone-bill-gates-facebook-melnet-speech-generation.

[8]"Lyrebird" is the system realized by the start-up, on the basis of deep learning models developed by the University of Montréal https://www.nextnature.net/2017/05/lyrebird-api-copies-human-voice/.

if AI solutions are based on rule sets, people have the opportunity of using abstract thought (Hadley 2019).

Finally, delegating cybersecurity to machines and replacing humans in control activities can certainly solve technical vulnerabilities, but if people and technology become ever more distant, other critical issues will have to be handled.

## 3.4 For a Holistic Vision of Cybersecurity

As discussed in the previous chapters, we have to get used to living in a more and more connected and digitized world, since this trend will continue in the years to come. Consequently, the amount of data available on the Internet will tend to rise, as well as the need to protect them.

Hence, it is realistic to think that the threat landscape will get worse.[9]

Individuals and organizations will be exposed to new security risks, and they will have to improve their capabilities to handle them. Indeed, criminals will continue to exploit any possible vulnerability, whether it is technological, physical or human.

Considering the current and the potential future situation of cybersecurity, a strong and effective approach to the issue becomes vital for everyone and, above all, for decision makers.

The first thing to do is to admit that, as it is, cybersecurity does not work. Like in a therapeutic relationship, recognizing the problem is fundamental to achieve positive outcomes.

Critical thinking is needed to develop alternatives to improve the current approach. We can start from the enhancement of what has been done so far, recognizing errors and weaknesses, and move on. It takes a change of mentality, involving both decision makers and those who have to deal with security problems. For example, within organizations, we should have the courage of breaking away from old patterns that consider security as a set of products or, worse, a check list to tick off.

Critical elements in the approach to cybersecurity cannot be ignored anymore, and we urgently need to redefine it.

Since the digital aspect is now an integral part of our lives, we should accept the idea of considering cybersecurity as a "public good" (Mulligan and Schneider 2011), like public health. In this sense, cybersecurity should be handled in the public interest, by developing a strong cooperation between public and private sector, as well as users' responsibility regarding their cybersecurity awareness (Taddeo 2019).

So, at least three issues are important in this vision.

---

[9]Just think of the COVID-19 pandemic which probably is going to change our future habits and way of working: increasing the dependence on digital tools exposes to the risk of cyberattacks (WEF 2020).

### 3.4.1 Excessive Focus on Technology

The first issue concerns the general approach to security, based mainly on a technocentric point of view. We confirm the importance of technological view to clarify how cyberthreats work and what technological solutions are needed, but this is not enough. It has been proved that technological solutions are not sufficiently developed to respond to all the threats.

We need to adopt a multidisciplinary vision of cybersecurity. Other disciplines—also far from technical approaches—are capable of giving different views. Clearly, these perspectives should be integrated with the technological approach, because we are strongly convinced that humans and technology have to work together.

Similarly, the study of the human factor in cybersecurity requires different contributions from social science fields, such as psychology, sociology, anthropology, and so on. To best understand cyberspace and all criminal activities developing in this huge area, it is not sufficient to be trained in mathematics and engineering related approaches (Patterson and Winston-Proctor 2019), but it is necessary the expertise of individuals with knowledge in behavioural sciences.

On the other hand, several international security reports over the years[10] have shown how the most advanced technological solutions are not capable of solving all cybersecurity problems. They represent only one of the means, not the only one.

There is a general overconfidence in emerging technologies as soon as they make their appearance, when instead they should not be seen as the magic bullet. Before relying on technologies completely, we should study their advantages and disadvantages.

On the contrary, it happens that technologies are immediately released on the market, while industry and media underline the pros of them. Unfortunately, the cons come after, when it is impossible to take a step back. Business is business, we know, and technological evolution must go on, but the basic question is: are we really willing to give up our ethical and social values for the benefits deriving from the use of technologies?

Differently from the past, we are now dealing with powerful technologies, but we are not sure that we will be able to control it, and this represents a serious problem for security too.

Other daily examples help us to understand how technology alone cannot offer a real solution to security. Besides a refined design, modern smartphones are becoming more and more equipped with highly advanced functions, such as sophisticated access to the device (fingerprint, facial recognition, etc.). Just pronouncing them, it seems that these technologies are able to keep hackers and criminals of all kinds away. However, being technologically advanced does not mean being protected from security risks. As security experts love repeating (and it is really true), an absolute security does not exist.

[10] See, for example, ENISA Threat Landscape Report (2019) and Verizon Data Breach Investigations Report (2017, 2018, 2019).

Nothing is impenetrable, considering that even a child can unintentionally overcome more or less sophisticated security protections.[11] Sometimes, people's false security perceptions are able to produce undesirable effects. Everyone knows the dramatic epilogue of the Titanic, marketed as unsinkable. Despite this assumption, things went differently, and unfortunately, consequences were tragic.

### 3.4.2 *Physical Elements Are Neglected*

The second issue concerns underestimating the "physical" elements involved in cybersecurity. Behind an attack there are devices and human beings, not only Internet. For example, attacks can guarantee the access to national infrastructures, with the consequence of interrupting essential services, such as transport, energy and financial systems. Coordinated cyber-physical attacks on critical infrastructures can be devastating and produce severe damages (Xiang et al. 2017). The well-known cyber-attack to Ukrainian power system in 2015[12] has shown that multiple approaches can be extremely effective, impacting remote assets both electronically and physically (Lee et al. 2016).

Then, thinking of Internet of Things, we know that this is made up of physical objects and connectivity; we already have evidence that the combination of the two elements can be critical for security.[13] We should not forget that even cities are implementing the adoption of smart technologies (smart cities), developing automation, remotely managed, and so on.

Looking at this scenario, we should realize that if the focus of cybersecurity remains anchored only to its cyber aspects, it is plausible to think that physical elements will be considered by cybercriminals as an attractive vulnerability to be exploited. It is no coincidence that criminal actions involve physical devices and that among criminal actions there are also theft and cards skimmer (Verizon 2018).[14] In

---

[11]See, for example, the news about a child who, at the age of 10, overcame his mother's i-Phone security block, consisting of the Face ID, the technology based on the recognition of the user's face https://www.wired.com/story/10-year-old-face-id-unlocks-mothers-iphone-x/. The episode is worrying, since if a child can unintentionally overcome security blocks, we can imagine what hackers can do.

[12]The Ukrainian power system cyberattack is the first publicly acknowledged incident to result in power outages. It left about 225,000 people without power for several hours.

[13]In 2016, a massive DDoS attack (distributed denial of service) hit the Internet. Important website platforms, like Twitter, Netflix, etc., took down. The company attacked was Dyn, that controls much of the internet's DNS infrastructure. The attack was realized through a specific malware (Mirai), which infected IoT devices (like DVR players, digital cameras), and accessing the devices using default password and usernames.

[14]The report refers to Payment Card Skinner, including all incidents in which a skimming device is physically implanted (tampering) on an asset that reads magnetic stripe data from a payment card.

addition, whether data is stored in a physical or digital form, physical access controls are required to prevent unauthorized access.

An effective approach to cybersecurity should involve actors and instruments coming from different security fields, together with people having a plurality of experiences in order to expand the breadth of analysis. On the other hand, analysing the new threats and the growing ability of criminals to exploit any vulnerability, it is evident that a traditional approach based on the separation of physical and cyber security is inadequate and should be overcome, since physical security regards the protection of personnel and IT infrastructures—hence hardware, software and data—from physical actions that could cause damages or loss to an organization.

Perhaps, speaking of Cyber-Physical-Security instead of just Cybsersecurity could be a more appropriated approach to understand and manage threats. Even though the term cybersecurity is more fashionable and appealing, we should not forget the relevance of physical elements involved.

### *3.4.3 Human Factors and Cybersecurity Culture*

The third issue—strictly related to the others—regards the importance of social and human aspects of cybersecurity, and the need of raising awareness. The building of Cybersecurity Culture is a must for any organization. In this sense, the popular triad "People, Technology and Process" developed by the security expert Schneier at the end of last century is still a valid reference to handle security.

Between the cyber and the physical dimension there are always human beings: whether it is a physical theft or a data breach, the illegal action is carried out by individuals, and the consequences inevitably involve them. Furthermore, people can also be the vector of criminal actions, such as in phishing and social engineering attacks, as discussed in this chapter.

In a new approach to cybersecurity, we should not see human factor as a problem, but as a part of the solution, since individuals are essential to the functioning of the socio-technical system (Zimmermann and Renaud 2019).

While everyone agrees that the human factor is fundamental, in practice this conviction does not find application as a global strategy, and when it happens, its management is not effective. We must admit that it is not a simple matter, and in designing cybersecurity awareness programmes we have to consider the need of taking care of cognitive, emotional and social aspects.

What is necessary is the construction of high-reliability-organizations (Winnefeld et al. 2015) based on interconnected and fundamental principles: integrity, depth of knowledge, procedural compliance, forceful backup, a questioning attitude, and formality in communication.

We often forget that organizations are made up of people, who must be prepared to recognize risks and respond to them, but they have to be put in the right condition to do so. Hence, investing in people also means paying attention to their well-being, since effective Cybersecurity Culture requires a receptive and a healthy workplace (Sect. 4.4).

Finally, working on prevention is still the best strategy. This concept is too often forgotten in favour of emergency management. Cybersecurity incidents happen constantly and it appears more and more difficult to prevent them, to the point where many are convinced that the only thing we can do is to be prepared to manage crisis events. We are aware that there are real hurdles in managing cybersecurity, since it touches every business process and function, and changing user behaviour requires considerable effort.[15]

On the contrary, we strongly believe that preparing employees to secure behaviour can prevent many security problems. After all, prevention is better (and cheaper) than cure.

## References

Aldawood, H., Skinner, G.: Reviewing cyber security social engineering training and awareness programs-pitfalls and ongoing issues. Future Internet, Review **11**(3), Art no. 73, (2019)

Bogen, M., Rieke, A.: Help Wanted: An Examination of Hiring Algorithms, Equity, and Bias, Upturn (December 2018)

Brundage, M., Avin, S., Clark, J., et al.: The malicious use of artificial intelligence: forecasting, prevention, and mitigation (2018)

Capgemini Research Institute: Reinventing Cybersecurity with Artificial Intelligence. The New Frontier in Digital Security (2019)

Cialdini, R.B.: Influence. The Psychology of Persuasion. Quill William Morrow and Company, New York (1984)

Cialdini, R.B.: Influence: Science and Practice, 4th edn. Allyn and Bacon, Boston (2000)

Cybenko, G., Giani, A., Thompson, P.: Cognitive hacking. Adv. Comput. **60**, 36–73 (2004)

CISCO: Annual Cybersecurity News Report (2018). https://www.cisco.com/c/dam/m/hu_hu/campaigns/security-hub/pdf/acr-2018.pdf

Corradini, I.: Human factors in hybrid threats: the need of an integrated view. In: Hybrid Cyber Warfare and The Evolution of Aerospace Power: risks and opportunities, CESMA (2017)

Corradini, I., Nardelli, E.: Social engineering and the value of data: the need of specific awareness programs. In: Ahram, T., Karwowski, W. (eds.) Advances in Human Factors in Cybersecurity, AHFE 2019. Advances in Intelligent Systems and Computing, vol. 960. Springer, Cham (2020)

Del Ponte, L.: European Artificial Intelligence Leadership, the Path for an Integrated Vision. Policy Department for Economic, Scientific and Quality of Life Policies, European Parliament, Brussels (2018)

Dhamija, R., Tygar, J.D., Hearst, M.A.: Why phishing works. In: Proceedings of the 2006 Conference on Human Factors in Computing Systems (CHI), Montreal, Quebec, Canada, pp. 581–590. ACM (2006)

ENISA: Threat Landscape Report 2018. 15 Cyberthreats and Trends (January 2019) https://www.enisa.europa.eu/publications/enisa-threat-landscape-report-2018

European Commission: Ethics Guidelines for Trustworthy AI (2019). https://ec.europa.eu/digital-single-market/en/news/ethics-guidelines-trustworthy-ai

[15]https://www.mckinsey.com/business-functions/mckinsey-digital/our-insights/why-senior-leaders-are-the-front-line-against-cyberattacks.

Europol: The Internet Organised Crime Threat Assessment (IOCTA) (2019)

Gupta, B., Arachchilage, N.A., Psannis, K.E.: Defending against phishing attacks: taxonomy of methods, current issues and future directions. Telecommun. Syst., 1–21 (2017)

Hadley, J.: In the Age of AI, The Human Factor Still Matters For Cybersecurity, Forbes, March 27 (2019). https://www.forbes.com/sites/jameshadley/2019/03/27/in-the-age-of-ai-the-human-factor-still-matters-for-cybersecurity/#7a9774725cc5

Hatfield, J.M.: Social engineering in cybersecurity: the evolution of a concept. Comput. Secur. (2017)

Houghton, E., Green, M.: People Analytics: Driving Business Performance with People Data. Chartered Institute for Personnel Development (CIPD), Global research, report (June 2018). https://www.cipd.co.uk/Images/people-analytics-report_tcm18-43755.pdf

Irani, D., Balduzzi, M., Balzarotti, D., Kirda, E., Pu, C.: Reverse social engineering attacks in online social networks. In: Detection of Intrusions and Malware, and Vulnerability Assessment, pp. 55–74 (2011)

Jansen, J., Leukfeldt, R.: How people help fraudsters steal their money: an analysis of 600 online banking fraud cases. In: Proceedings of the 5th Workshop on Socio- Technical Aspects in Security and Trust, pp. 25–31 (2015)

Krombholz, K., Hobel, H., Huber, M., Weippl, E.: Advanced social engineering attacks. J. Inf. Secur. Appl. **22**, 113–122 (2015)

Kruglanski, A.W., Thomson, E.P.: Persuasion by a single route: a view from the unimodal. Psychol. Inq. **10**, 83–109 (1999)

Lee, R.M., Assante, M.J., Conway, T.: Analysis of the Cyber Attack on the Ukrainian Power Grid. SANS Industrial Control Systems, Santa Monica, CA, USA (2016)

Mitnick, K.D., Simon, W.L.: The Art of Deception: Controlling the Human Element of Security. Wiley Publishing, Inc., Indianapolis, IN (2002)

Mouton, F., Leenen, L., Venter, H.S.: Social engineering attack examples, templates and scenarios. Comput. Secur. **59**, 186–209 (2016)

Mulligan, D.K., Schneider, F.B.: Doctrine for cybersecurity. Daedalus **140**(4), 70–92 (2011)

Patterson, W., Winston-Proctor, C.E.: Behavioural Cybersecurity: Application of Personality Psychology and Computer Science. CRC Press, Taylor & Francis (2019)

Perloff, R.M.: The Dynamics of Persuasion. Communication and Attitudes in the 21st Century, 5th edn. Routledge (2014)

Petty, R.E., Cacioppo, J.T.: The elaboration likelihood model of persuasion. In: Berkowitz, L. (ed.), Advanced in Experimental Social Psychology, vol. 19, pp. 123–205 (1986)

Salahdine, F., Kaabouch, N.: Social engineering attacks: a survey. Futur. Internet **11**(4), 89 (2019)

Segovia, L., Torres, F., Rosillo, M., Tapia, E., Albarado, F., Saltos, D.: Social engineering as an attack vector for ransomware. In: Proceedings of the Conference on Electrical Engineering and Information Communication Technology, Pucon, Chile, 18–20 October 2017, pp. 1–6 (2017)

Taddeo, M.: Is cybersecurity a public good? Mind. Mach. **29**(3), 349–354 (2019)

Verizon: Data Breach Investigations Report (DBIR) (2017). https://enterprise.verizon.com/resources/reports/2017_dbir.pdf

Verizon: Data Breach Investigations Report (DBIR) (2018). https://enterprise.verizon.com/resources/reports/DBIR_2018_Report.pdf

Verizon: Data Breach Investigations Report (DBIR) (2019). https://enterprise.verizon.com/resources/reports/dbir/

WEF: Why Cybersecurity Matters more than ever During the Coronavirus Pandemic (March 2020). https://www.weforum.org/agenda/2020/03/coronavirus-pandemic-cybersecurity/

Winnefeld, J., Kirckhoff, C., Upton, D.A.: Cybersecurity's human factor: lessons from the Pentagon. Harv. Bus. Rev., 87–95 (September 2015)

Xiang, Y., Wang, L., Liu, N.: Coordinated attacks on electric power systems in a cyber- physical environment. Electr. Power Syst. Res. **149**, 156–168 (2017)

Zimmermann, V., Renaud, K.: Moving from a "Human-as-Problem" to a "Human-as-Solution" cybersecurity mindset. Int. J. Hum. Comput. Stud. **131**, 169–187 (2019)

# Chapter 4
# Building a Cybersecurity Culture

**Abstract** Cybersecurity Culture (CSC) is unique for an organization and not replicable for others, being based on its specific characteristics, in terms of technologies, processes, and people's values. To be effective CSC has to be considered as an integral part of the Organizational Culture (OC), since it can reinforce the cultural cohesion within organizations and lead people to realize that security is not just an IT department problem, but an issue which involves everyone. The building of CSC requires adopting a clear strategy and a set of resources, methods and tools. The premise is that a cooperative approach is better than imposing procedures, given that people do not often follow rules which are imposed and, above all, that they do not understand. The first and foremost issue for building CSC in an organization is an assessment phase, an internal "diagnosis" in order to evaluate the current state of the maturity of its security culture. This step is essential to avoid adopting security awareness programmes which do not meet the actual needs of the organization. As a good recipe depends on its ingredients, an effective CSC depends mainly on the active commitment of people. In this chapter implications of the construction of a CSC are discussed, together with the main framework and tools available in literature; moreover, some simple recommendations, such as the real involvement of top management, are provided in order to steer the work in the right way. Finally, to achieve positive outcomes, we should ensure a receptive and healthy environment; we can assume that when people perceive a safe work environment, they are more likely to carry out their tasks and to adopt a secure behaviour.

## 4.1 Organizational Culture and Cybersecurity Culture

Before discussing the fundamental elements for the construction of a Cybersecurity Culture (CSC), we have to understand its connection with Organizational Culture (OC), a topic which has been widely studied in literature.

People are the key factor of OC: if culture is "in the mind and heart of men" (Goudenough 1956) and it can be defined as "software of the mind" (Hofstede et al. 2010), then OC collects people's ideas, beliefs, behaviours and other elements that ensure the survival of the organization.

© Springer Nature Switzerland AG 2020

I. Corradini, *Building a Cybersecurity Culture in Organizations*,
Studies in Systems, Decision and Control 284,
https://doi.org/10.1007/978-3-030-43999-6_4

**Table 4.1** Some definitions of Organizational Culture

| |
|---|
| The pattern of basic assumptions that a given group has invented, discovered, or developed in learning to cope with its problems of external adaptation and internal integration, and that has worked well enough to be considered valid, and therefore, to be taught to new members as the correct way to perceive, think and feel in relation to these problems (Schein 1985) |
| The collective programming of the mind which distinguishes the members of one human group from another (Hofstede 1984) |
| The pattern of shared beliefs and values that give members of an institution meaning, and provide them with the rules for behaviour in their organization (Davis 1984) |
| The glue that holds an organization together through a sharing of patterns of meaning. The culture focuses on the values, beliefs, and expectations that members come to share (Siehl and Martin 1984) |

The concept of OC calls on a wider notion of culture, studied mainly by anthropologists, like a collective way of thinking and acting. Focusing on organizational culture and its connection with security, we cannot ignore the influence of national cultures on people's values and beliefs, and consequently on individuals' attitudes toward security.

Indeed, despite the abstract nature of culture, it can affect individual and organizational behaviour (Cameron and Quinn 2011). For instance, culture can have an effect on decision making relative to computer security risks (Chen and Farkas 2009), as well as being a significant factor for predicting attitude to privacy (Halevi et al. 2017).

Lots of definitions on OC are available in literature (Table 4.1), but in short it can be identified as "the way we do things around here" (Lundy and Cowling 1996).

Analysing the different perspectives provided by authors, it is clear that the concept of OC includes values, cohesion, sense of belonging, attitude and behaviour. It is equally clear that it has several important functions within organizations (Schein 1985; Martin and Siehl 1983), since it encourages the cooperation between members, defines an organizational identity, sets out strategies and goals, shapes attitudes and behaviours.

OC is based on cultural norms defining what is accepted or not; therefore, when it is aligned with personal values and needs, it is able to develop a huge energy and help foster organizations to thrive (Groysberg et al. 2018). From a business perspective, organizations with a strong culture have a better performance and show a competitive advantage (Barney 1986). In fact, thinking of the most important and successful organizations in the world, we observe that they have distinctive organisational cultures which make them unique and easily recognizable.

Together with the dominant culture, other sub-cultures can coexist in an organization (in particular in highly structured ones); this is not a negative condition, because when these sub-cultures are consistent with the dominant culture, they can even reinforce it.[1]

[1] Differently from sub-cultures, counter cultures do not accept the dominant culture and express opposition, not useful for the success of the organization.

OC has some "traits", just like individuals have their personality. Consequently, understanding OC is equivalent to understanding the personality of the organization. We can see the exterior aspects of a company, but its interiority is not visible to the naked eye. If we want to find out what happens under the surface, we need to proceed with an internal investigation.

The concept of OC is frequently described with the iceberg metaphor, where the visible elements are on the surface, while below there are the hidden ones.

The visible elements are made up of artifacts, consisting of the tangible level of an organization, for example the physical environment, the technological tools used by employees, individual behaviour.[2] Even if clearly visible, it does not mean that artifacts are easily decodable.

In the hidden level there are the values (e.g. beliefs, implicit and explicit norms) spread over and shared within organizations to enhance a sense of belonging, the clarification of organization choices, and so on.

At the deepest level there are tacit assumptions and unspoken beliefs taken for granted, representing the true essence of an organization culture (Schein 1999). Assumptions are related to universal human experience, like the relationship with nature or the sense of time.

To comprehend the relationship between CSC and OC, it is helpful to define what CSC is, since its definition includes a good deal of factors of the other concept, with the addition of the aspects of security and technology. According to one of the best know definitions, CSC refers (ENISA 2018) to

> the knowledge, beliefs, perceptions, attitudes, assumptions, norms and values of people regarding cybersecurity and how they manifest in people's behaviour with information technologies.

Hence, it is not just a matter of protecting information, but something more: CSC is understood in a broad perspective as an integral part of people's lives.

Considering what discussed above, CSC is made up of different elements belonging to the different levels of the OC model, including—among others—awareness training, papers describing the values and the principles of the organization, the way employees interact with technological devices, and so on (see Table 4.2).

Besides the three levels of Schein's model, a further element can be considered, represented by information security knowledge (Van Niekerke and Von Solms 2010). This fourth level supports the others, given that employees' knowledge fosters people to understand the relevance of certain rules, functions and behaviours.

The importance of security culture in an organization has widely been discussed in relation to the protection of information (Dhillon 1995; Hinson 2003; Kearney 2010; Schneier 2000; Schultz 2005), bringing out how a strong relationship between OC and security generates positive results. In particular, the relationship

---

[2]Hatch (1997), explaining artifacts, distinguishes physical manifestations (e.g. buildings, logo), behavioural manifestations (e.g. rituals, ceremonies) and verbal manifestations (e.g. anecdotes, stories).

**Table 4.2** Elements of CSC

| Cybersecurity culture | Examples | In practice |
|---|---|---|
| Artifacts | Awareness training<br>Technological devices<br>Employees' behaviour | What we see |
| Values | Documents and papers describing values and principles of the organization<br>Rules of conduct | What we share |
| Basic assumptions | Sense of time connected to the idea of progress and advancement in technology<br>Consideration on the human contribution to work environment | What we think and believe |

between Information Security Culture (ISC)[3] and OC can be decisive for developing responsible and attentive people's behaviours (Fitzgerald 2007; Lim et al. 2009).

Investigating this connection, it emerges that when ISC is embedded into OC, the relationship between the two is strong and so are motivation and awareness; moreover, protective behaviours are adopted by employees. On the contrary, when ISC is a subculture of OC or separated from it, risk vulnerabilities grow, and responsibility is not felt like a priority for employees.

The starting point is that CSC has to be considered as an integral part of the OC, since technologies and processes are strictly related in the modern organizations. In this way it is possible to reinforce the cultural cohesion relative to the security aspects; the higher the consensus, the more organizations strengthen their security culture. Achieving this goal requires changing people's attitude towards security: this is the real challenge given that for those who do not deal with this issue, security is often perceived as an unprofitable workload.

Security culture should be part of the way of thinking and behaving of all employees, and not something external or to delegate. Considering the CSC definition and the large use of digital technologies both in professional and in private life, we should not limit our interest to working behaviour, but to extend as much as possible the range to people's private life and their home environment (De Veiga 2019). In the current information society, we manage information at any time and in different ways; consequently, information security awareness should be an integral part of the general knowledge of citizens (Siponen 2001).

This is why cybersecurity should not be addressed only to IT experts, but it should involve everyone: regardless of role, everybody needs to know how to move with confidence in the digital environment.

[3]Despite the differences and similarities emerged in literature between the concept of Information Security Culture and Cybersecurity Culture, in this book these terms are used interchangeably for the reason discussed in Chap. 1, Sect. 1.2.

## 4.2 The Premise

For the development of an effective Cybersecurity Culture we need to consider all aspects discussed in the previous chapters, including the need of adopting a holistic approach (Chap. 3, Sect. 3.4).

The premise is that promoting a CSC is better than imposing rules and policies, since people do not easily follow rules which are imposed. They can comply with security policies of the organization, but if they are not aware of the importance of their actions, we cannot be sure that the adherence to these rules produces positive results.

Encouraging people instead of forcing them to do or not to do some actions can achieve better outcomes. Moreover, when something unexpected occurs and procedures are not enough to solve the problem, individuals might feel inspired to think of alternative solutions.

Handling a cultural change and building a strong Cybersecurity Culture is not a simple process and it requires time, resources, tools and, above all, an active support from top management. We cannot expect immediate results because people are not machines programmed to passively accept all they are told to do.

In the era of Artificial Intelligence, someone might think of relying on algorithms to drive people behaviour, convinced that individuals can be governed by machines and algorithms. But beliefs, values and emotions are an integral part of human mind, hardly replicable, and if we want a real process of cultural change, we must invest in people. We have to avoid choosing short-cuts, because the goal we want to achieve is ambitious and too easy ways out may be a failure. For people data protection should become a natural way of managing their daily work (Oost and Chew 2007; Siponen 2000).

Two types of behaviour derive from an effective Cybersecurity Culture (Huang and Pearlson 2019): actions connected to the specific role in the organization, and other useful activities not strictly related to the job description, like cooperative behaviours and offering comments to improve cybersecurity.

Another important issue is that CSC is not just for large companies. Since cybersecurity is a priority for all countries, it is evident that organizations, from small enterprises to large corporations, must to be ready to face cyberthreats. Indeed, we must not forget that security is a strategic asset for business and for companies reputation (Chap. 5, Sect. 5.3).

Before developing a process of change, organizations have to be clear how much they are willing to invest in cybersecurity education, and what their expectations are. For instance, we cannot think that a one-off training course is able to transform people's mentality. If this change were so easy, the security problem related to the human factor would have already been settled.

A training course can certainly instil in employees the idea of the importance of human behaviour in security, but changing their actual behaviour is another question. This consideration can be taken for granted, but in practice companies expect to achieve results in a short time, while CSC requires to be constantly nourished.

We have to think of CSC as a step-by-step process and to identify clear questions, whose answers depend on the security maturity level of our organization. Some questions can help our process, for instance: What is the initial condition from which to start? What do we want to achieve? What methods and tools do we need to implement security awareness? How to measure results?

The construction of CSC is a process aiming at a continuous quality improvement, based on a logical sequence of four steps (Plan-Do-Check-Act),[4] known also as "Deming Cycle" in the process management field.

Improvement has to involve of all the elements of the process, including human beings. Soliciting employees' feedback on security policies or security initiatives is an example of monitoring action able to provide useful information for improving the process. CSC is an ongoing process which can be effective only if it is constantly monitored.

## 4.3 An On-Going Process

Cybersecurity Culture is unique for an organization and not replicable for others, being based on its specific characteristics in terms of technologies, processes, and people's values. Hence, the design of cybersecurity awareness programmes has to take into account all these elements in order to be contextualized and to fit with the specific needs and expectations of the company.

For a better management of the process, it can be useful the creation of a working group (a Cybersecurity Culture team) which should guide, coordinate and monitor the execution of activities. It is vital this team receives its appointment directly from top management, so as to signal to the entire company the importance of its action.

Obviously, the presence of this group and its composition depends on the size and the type of the company. Indeed, in large companies the group should count on the presence of representative members of the different areas of the organization, like Human Resources, IT Security, Legal, Risk, Compliance, Communication (ENISA 2018). This approach is not applicable in a small company or perhaps in a medium one; however, it is equally important to identity who is going to have the role of monitoring all security awareness activities.

The first and foremost issue for building CSC in an organization is an assessment phase, an internal "diagnosis" in order to evaluate the current state of the maturity of its security culture. This step is essential to avoid adopting security awareness programmes which do not meet the actual needs of the organization.

The assessment should involve the key people of the company (Wilson and Hash 2003), like executive management, security personnel, and adopt a combination

[4]This philosophy of work characterizes several ISO standards, like the international standard ISO 27001 for IT security (https://www.itgovernance.co.uk/iso27001), whose goal is to ensure right policies and procedures within organizations to manage their information assets. This standard emphasizes the importance of risk management.

of tools (e.g. questionnaires and interviews) to gather information useful to plan adequate activities.

Lots of theoretical frameworks and models are available in literature to identify the main steps and methods for an assessment of security culture, and some of them will be discussed in the subsequent paragraph.

On the basis of the literature (e.g. Schlienger and Teufel 2003; Martins and Eloff 2002; Niekerk and Von Solms 2010) we summarize stages and activities (Table 4.3) supported by examples of the questions that an organization (and specifically its Cybersecurity Culture team) should pose to guide the process.

Starting from an assessment phase and developing activities to raise security awareness, companies have the opportunity to identify their strengths and weaknesses and to define the process and the actions more suitable for them.

**Table 4.3** Essential stages for developing a cybersecurity culture

| Phases/Activities | Objectives | Questions (examples) |
|---|---|---|
| Preliminary activity | Prepare the organization to receive cybersecurity security programmes | What has been done up until now?<br>What has worked or not, and why?<br>Does top management support the process?<br>Who is going to handle the activities? |
| Assessment phase of security maturity | Identify the level of security maturity of the organization (diagnosis) | What is the current state of Cybersecurity Culture?<br>What are the critical and positive elements? |
| Planning cybersecurity awareness/training programmes | Develop tailor-made programmes according to the different level of organizations and considering the results of the diagnosis<br>Consider the need of a continuous training | What tools and methods are better for our audience?<br>Do we own the proper skills to design effective training courses? |
| Communication | Develop an adequate communication plan to support the activities | What tools are better to support cybersecurity awareness programmes? |
| Monitoring | Monitor the process and promote adjustments | Is there something missing or that can be improved? |
| Measuring the results | Evaluate the impact of security awareness programmes | How to evaluate the effectiveness of cybersecurity awareness programmes? |
| Further deployment | Identify further activities deriving from the analysis of results | What can we do to maintain/improve our security approach? |

Stages do not necessarily have to be interpreted as a flow diagram; we prefer to consider these elements as something that organizations should include according to their security maturity level. For instance, a company which has never developed security programmes has different needs compared to a company which has already undertaken a security path. In the same way, a small enterprise has different needs compared to a large one. However, the premise is that security culture has to be aligned with the organization; therefore, it requires being updated as long as the organization changes to adapt to the evolving operating scenario.

It is therefore important to carefully prepare the organization for a different approach to security. This stage precedes the assessment phase, since its goal is to evaluate what company has done in the past and up until now.

Currently, several frameworks for security culture have been developed especially for large companies. However, we know that the economic fabric of a nation is made up of all kinds of companies, and that in an interconnected and global environment any firm, whatever its dimension, can become a vulnerability for the entire system. All companies, even the smallest, are called to use appropriate measures to protect their business.

In addition, regardless of the cybersecurity framework adopted, the building of a CSC requires constant adaptation and deployment, considering the changes within organizations (including staff renewal) and the new threats to be faced.

In designing the process, it is fundamental to identify the best tools and methods for security awareness and training (Chaps. 6 and 7) and how to measure the outcomes, considering that training initiatives positively affect the building of a security culture (Martins and Da Veiga 2015).

Given the specialized activities and the need of ensuring an accurate process, companies have to ask themselves if they own the right skills to use the proper methods and techniques to identify their level of security maturity, as to design tailor-made training programmes.

Finally, the assessment process is much more than the administration of questionnaires/interviews to employees, just like communication has to be evaluated as a strategic resource to support all activities, and not only to announce the new initiative on cybersecurity that company aims to promote.

## 4.4 Approaches for the Assessment

Many empirical studies have defined some useful approaches aiming at identifying the main components of a security maturity assessment. Even though the importance of fostering a security culture within organizations is a widely shared objective, how to conduct the assessment is still a critical task: none of the different approaches proposed seem to be capable of investigating all levels of information security culture (Okere et al. 2012).

However, frameworks, models, tools and perspectives available are useful resources for companies in order to identify the dimensions to be investigated and how to study them for implementing specific security awareness programmes, according to their other organizational factors.

### 4.4.1 The Organizational Behaviour Model

One of the most critical issues is to integrate the human factor into the organizational context for managing cybersecurity. In fact, even if everyone agrees on the importance of this step, it is not easy to find a model able to include technological elements and people within organizations. An approach based on organizational behaviour[5]can respond to this need.

The following model (Schlienger 2006; Schlienger and Teufel 2003)—also discussed by ENISA in its Guidelines on Cybersecurity Culture (2019)—represents an interesting approach to gaining awareness, since it looks at the different levels of an organization. Starting from the consideration that security culture is a never-ending process and that it has to correspond with the targets of the organization, the preliminary phase of assessment is fundamental to evaluate this correspondence.

The assessment can be done through a combination of different tools (Schlienger and Teufel 2005) which provide a global view of the organization. For instance, using a standardized questionnaire realized on the basis of organizational behaviour model (Robbins 2001), it is possible to investigate the three levels of organizational behaviour—organization, group and individual (Table 4.4)—and their interconnections.

The strong point of this approach is represented by the relevance of both individual factors (e.g. perception, motivation) and the organizational ones (e.g. organisational culture and structure) for the assessment of CSC.

Furthermore, this view underlines the relevance of practical activities, like communication and training: not only the lack of training, but other factors can be critical and affect the work environment, like problem management and work and technology design.

**Table 4.4** The three levels of organizational behavior

| Individual | Group | Organization |
|---|---|---|
| Values | Leadership | Work and technology design |
| Perception | Communication | Organizational culture |
| Attitude | Problem management | Organizational structure |
| Motivation | – | Training |
| Knowledge | – | – |

[5]Organizational behavior is a wide area of study based on interdisciplinary view to investigate how individuals and groups interact within organizations and towards them.

**Table 4.5** Security dimensions in ISFC model

| Dimensions | examples |
|---|---|
| Strategy | Policies, best practices, guidelines |
| Technology | Hardware, software, applications |
| Organization | Beliefs, values, norms |
| People | Individuals behaviour within organization |
| Environment | National culture, regulatory systems, government initiatives |

### *4.4.2 The Information Security Culture Framework (ISFC)*

This approach (Hogail 2015) proposes an integration between human beings, technology and organization, with the goal of providing a comprehensive framework able to assess information security culture in organizations.

Five dimensions are identified in the framework: strategy, technology, organization, people, environment and change management (STOPE), and each one includes tasks covering four significative domains of human behaviour (preparedness, responsibility, management, society and regulations). In this way it is possible to analyse the relationship between the different factors involved (Table 4.5).

The five dimensions are investigated through a questionnaire, including demographical information, and semi-structured interviews, whose goals are to collect information concerning people's assumption and behaviour relative to information security. The model incorporates change management principles (e.g. motivation, communication, sufficiency of resources) to guide the development of information security culture.

Finally, the model considers the relevance of external elements (the environment) in the process: it is evident that national regulations and initiatives on cybersecurity shape the strategies within organizations.

### *4.4.3 The Competing Security Cultures Framework (CSCF)*

This approach (Hayden 2015) supports organizations to identify all those elements representing a risk for their security goals. It is based on an adaptation of the Competing Values Framework (Quinn and Rohrbaugh 1983), one of the most used models for research on organizational culture, initially developed with the purpose of identifying indicators of organizational effectiveness.

With the goal of measuring those traits which can strengthen or weaken security performance, the CSCF model adopts the quadrant structure of the original framework, reconfiguring it into a people-centric security model. Each quadrant represents a separate culture approach—Process Culture, Compliance Culture, Trust Culture and Autonomy Culture—and consists of different groups of values (e.g.

**Table 4.6** The ten areas investigated in the security culture diagnostic survey, SCDS

| |
|---|
| What is valued most? |
| How does the organization work? |
| What does security mean? |
| How is information managed and controlled? |
| How are operations managed? |
| How is technology managed? |
| How are people managed? |
| How is risk managed? |
| How is accountability? |
| How is performance evaluated? |

stability, flexibility, communication, conformity), affecting security decisions and activities developed within the company. Effectiveness depends on the degree of security control and on internal and external focus, determining if organization is primarily focused on security as a function of the organization itself or not. Despite the four distinct approaches, the author of the model underlines the combination of cultures characterizing organizations.

To investigate cultural traits and security values, a tool (The Security Culture Diagnostic Survey, SCDS) made up of ten areas to be investigated with more detailed questions (Table 4.6) has been proposed by the author of this framework.

### *4.4.4 The Information Security Culture Assessment (ISCA)*

Cultivating an effective security culture require focusing on employees' behaviour, and implementing necessary actions in order to achieve an acceptable information security culture.

The Information Security Culture Assessment—the ISCA questionnaire (Da Veiga and Eloff 2007, 2010)—is a validated statistical tool made up of the 10 dimensions (constructs) described in Table 4.7 (Da Veiga 2019). It can be used to monitor the information security culture and to identify areas where intervention is needed to improve employees' behaviour regarding the protection of information.

Analysing the different dimensions, it is possible to understand how acceptable the level of information security culture is for an organization. As the information security maturity level of each organization can be different, for effective outcomes it is necessary to verify that all questions are relevant to the working environment to be investigated. To understand how the constructs of ISCA positively affect information security culture, further investigations have been conducted. On the basis of these results, authors have developed the conceptual Information Security Culture Model (ISCM), providing important information for the management in order to take the

**Table 4.7** Dimensions investigated by ISCA questionnaire

| Dimensions | What to assess |
|---|---|
| Information asset management | User's perception about the protection of information assets |
| Information security management | Management's perception of information security management |
| Change management | Users' perception about changes and users' willingness to change in order to protect information |
| User management | Users' awareness and training for the protection of information |
| Information security policies | The understanding of information security policies by users and the effectiveness of its communication |
| Information security programme | The effectiveness of investing in information security resources |
| Trust | Users' trust in communication within organization and their perception of the safekeeping of private information |
| Information security leadership | Users' perception of information security governance |
| Training and awareness | Users' perception of additional needs for information security training |
| Privacy perception | Users' perception of privacy principles |

right direction for the proper handling of information security culture (Martins and Da Veiga 2015).

### *4.4.5 The Security Culture Toolkit*

This approach combines different perspectives to meet the security needs of the organization, and include psychological, sociological and anthropological points of view. According to this security culture model (Laycock et al. 2019) seven dimensions are highly regarded to reduce risks, because they are connected to three pillars of each organization: people, tools and processes (Table 4.8).

It is worth noting that communication is included among the dimensions listed in the model: this is an important step, which is too often overlooked. Instead, for the building of CSC, communication plays a significant role, since it supports all the activities. The model stresses the relevance of identifying security metrics to measure security culture in order to demonstrate the progress made. The authors assume that "you cannot improve what you do not measure". In this sense, the approach provides independent security culture metrics to measure and record the changes achieved.

**Table 4.8** Dimensions of the security culture toolkit

| Dimensions | Description |
|---|---|
| Attitude | The feelings and beliefs that employees have toward the security protocols and issues |
| Behaviours | The actions and activities of employees having an impact (direct or indirect) on the security of the organization |
| Cognition | The employees' understanding and their self-efficacy of security knowledge |
| Communication | The quality of communication and support on security matters |
| Compliance | Being adherent to rules, policies and regulations |
| Norms | The knowledge of an adherence by employees to unwritten rules (how security-related behaviours are perceived by them, as accepted or unaccepted) |
| Responsibilities | How employees perceive their role with respect to the organizational security |

### 4.4.6 *The Cyber-Risks Perception Analysis*

Another approach is to focus on employee's cyber-risk perception as an essential step to define the process of building CSC (Corradini and Nardelli 2016, 2019).[6]

This activity allows to evaluate how employees perceive security risks of their organization and within their private life, and how they respond to them. In a few words, it is possible to investigate how employees are sensitive and informed about cybersecurity risks. At the same time this gives a clear picture of the organization for the planning of the subsequent activities. To be effective, this analysis requires a combination of methods, such as questionnaires, interviews and focus groups; in this way it is possible to have a complete overview based on quantitative and qualitative analysis.

This approach, if designed and implemented on the basis of an accurate analysis of the current organizational culture, can develop a risk-aware culture, including cybersecurity risks. Considering the daily presence of digital technologies both in professional and in private life, cyber-risk perception should be investigated in both areas, analysing for example:

- individual factors affecting risk perception, its evaluation and its management;
- people's perception about information technologies and cyber-risks;
- how employees perceive risks in the organization and in their private life;
- what prevention measures and protective behaviours should be included in the management strategy of cyber-risks.

[6]This survey is part of a wide initiative called "Risk Culture Project", designed in cooperation with the Risk Management Team and Human Resource Team of a multinational company operating in the financial sector. The goal is to increase employees' awareness on the different risks, both implementing a continuous training and validating/updating the company's risk policy.

The idea of investigating risk perception in both areas is interesting to understand the coherence and the difference between how individuals perceive cyber-risk in private life and at work.

On the basis of the outcomes of the survey conducted within organizations, it is possible to design a cybersecurity awareness programme adapted to the specific needs of the company.

## 4.5 Recommendations

Some recommendations can help in enhancing the process of building a Cybersecurity Culture. Not all of them are applicable in the same way, because it depends on particular conditions, like the size and the structure of the company. For example, as discussed in previous paragraphs, in small businesses it is objectively impossible to count on the presence of cross groups for coordinating and monitoring the cybersecurity activities.

The basic philosophy is that security is a priority for everyone: in an interconnected business world even one company can be a Trojan horse for another. Nevertheless, cyber risk is not considered systematically by organizations in their supply chain, and few companies demand minimum standards from their suppliers (Cyber Security Breaches Survey 2019).

As a good recipe depends on its ingredients, an effective Cybersecurity Culture depends mainly on the active commitment of people. However, some recommendations are helpful to steer the work in the right way.

### *4.5.1 Real Involvement of Top Management*

Since cyber-threats are a business problem, the C-suite[7] management should be naturally interested in security issues, also giving a significant contribution to decisions about budget and business. Instead, cybersecurity is often seen as a task delegated to IT department and not a strategic asset for business.

The successful of security initiatives depends on the real involvement of top management, since it can have a significant direct and indirect influence on employees' attitude towards information security polices, as on organizational culture (Hu et al. 2012). When executives assume a behaviour complying with security policies, they become a role model for their personnel; the key message is that cybersecurity is an issue that affects the entire organization and that everyone has to do their part.

[7] In business language, C-suite denotes the group of the most important executives in a company, in particular those whose titles begin with the letter C, for chief (Cambridge Business English Dictionary). https://dictionary.cambridge.org/it/dizionario/inglese/c-suite.

It is therefore essential to involve top managers and make them aware of the importance of cybersecurity awareness programmes. At the same time, it is important that they themselves receive an adequate training in cybersecurity: they use digital technology for work and in their private life, and they are exposed to cyberthreats as well as their employees. They handle sensitive information and they travel frequently using different digital tools, thus becoming a priority target for cybercriminals. It is evident that they have to be the first to take a serious interest in the subject and set a good example.

### 4.5.2 *Different views and perspectives*

Facing cyber-threats requires the adoption of new approaches and different views. In this sense, "diversity" can be a source of enrichment.

According to the current research, people from diverse backgrounds might affect a group's behaviour, fostering a more accurate group thinking and are more willing to re-examine facts with a certain objectivity (Rock and Grant 2016). Looking at gender diversity, for instance, research shows that gender-diverse teams at the business unit level positively affect financial performance (Badal and Harter 2014), and that the presence of women in groups improves team collaboration (Bear and Woodley 2013).

Women can provide a different perspective on the analysis of the cybersecurity landscape, since this is more and more populated by criminal actors with different backgrounds aiming at searching for new vulnerabilities and opportunities to exploit. To contrast such a criminal approach, we need to think of creative perspectives and to find out-of-box thinkers for analysing security problems in an original way.

Moreover, we know that the multiple vision of the security problems is also due to a diverse way of gauging risks (O'Flaherty 2018). The prevailing narrative considers women more risk adverse than men; the fact that women and men perceive and respond to risk differently has to be interpreted as a stimulus to build mixed gender teams.[8]

Women participation in cybersecurity field is growing,[9] but there is still much to do. Finally, considering that diversity comes in many forms (e.g. age, gender, race) and produces measurable results (Eftekhari and Spaniel 2020), organizations should encourage dynamic teams.

---

[8]See, for example, the report Women in business: new perspectives on risk and reward by Granth Thornton, March 2017, https://www.grantthornton.co.uk/globalassets/1.-member-firms/united-kingdom/pdf/publication/women-in-business-new-perspectives-on-risk-and-reward.pdf.

[9]Regarding female participation in cybersecurity see, for example, "Women in Cyber Security Literature Review", by the Department of Prime Minister and Cabinet (PM&C) in cooperation with Australian Centre for Cyber Security of University of Canberra (Foley et al. 2017).

### 4.5.3 The "Cultural Threats"

For the success of cybersecurity initiatives, organizations should consider that other types of threat have to be handled. Deciding to implement Cybersecurity Culture requires considering all internal processes and dynamics, the so-called "covert processes" (Marshak 2006), responsible for the failure of organizational change initiatives. In this sense, blind spot or tacit assumptions can represent an invisible force able to have a strong (and negative) impact on the change process, slowing it down or blocking it.

In the PEPL model (Hayden 2015) the author describes four forms of threats (political, emotional, psychological and logistical) and the related cultural weaknesses, included their effects on security outcomes. In particular, cultural weaknesses consist of (for example):

- lack of controls or policy enforcement (for political threat);
- expectations of rational behaviour (for emotional threat);
- improper communications or training (for psychological threat);
- problems associated with new security objectives (for logistical threat).

It is not an easy thing to identify these cultural weaknesses; however, this activity is essential to handle all those obstacles to the building of a successful CSC.

Besides what discussed above, there are some myths related to cybersecurity that need to be overcome within organizations. For instance, hackers are not the only threat to face, since also insiders or any individual managing sensitive data is a potential menace. In the same way, it is not true that the more organization spend in security, the more secure they will be (Mc Kinsey 2019). The quality of the investment, as well as the construction of a trust environment, are fundamental to develop CSC.

### 4.5.4 People and Policies

One critical point concerns the employee adherence to company security policies, whose lack can generate significant consequences. Security policies are often perceived by employees as a further task, or worse, as a tool used by the company to exercise control over them. Security policies are fundamental for organizations, because they are expression of their organizational culture. They represent (NIST, Glossary)[10]:

> A set of rules that governs all aspects of security-relevant system and system element behaviour.

[10]According to this definition, system elements include technology, machine and human. https://csrc.nist.gov/glossary/term/security_policy.

Security policies are not something static, but dynamic documents that need to be updated, both in agreement with the evolution of threats and with organizational changes.

When individuals decide to comply or not with security rules, they carry out a cost-benefit analysis (Beautement et al. 2008). In this sense, some costs perceived by individuals, like physical or cognitive load, can lead them to give up following the rules. This problem does not have a simple solution. It is possible to adopt two different perspectives (Tyler 2005): an approach based on the punishment consequent to an undesired behaviour (command-and-control model) or one based on developing people internal motivation (self-regulatory model).

Regardless of the approach used, the important thing is to overcome employees' reluctance for rules and policies. Considering the goal of implementing CSC, we prefer to invest on people's motivation and the reinforcement of their ethical values. In fact, one the most important reasons explaining why employees are not engaging with security policies is that they are not fully aware of their organization IT security policies.[11] In many cases, they do not understand them because of the complex language, or they do not consider them relevant to perform their tasks.

In this sense, communication strategies play an important role. For instance, discussing security policies—in dedicated meeting or in informal way—together with employees has a different impact compared with other communication means, such as sending an email describing them or inviting people to consult the company intranet. We are aware that such an approach requires time and effort, especially in large organizations, but time spent with personnel is never wasted time. Where present, the cybersecurity team can be an important resource for this task. Moreover, interacting with employees on security issues improves their motivation, since they feel empowered and involved in an important organizational strategy. On the other hand, the acceptance of the company's policies, even the security ones, is part of work employees, but also of business strategy.

Another communication element is related to messages used to make individuals aware of the relevance of security policies. Rather than focusing just on the negative consequences of the failure to execute procedures, it would be useful to propose successful experiences. These can enhance people's perception about their positive role and reinforce their self-esteem. Underlying only the risks and the negative consequences of a non-adherence to policies can be counterproductive and lead people to remove this kind of communication.

Finally, since organizational culture can positively affect employees towards security policies, management should keep in mind this aspect when deploying new security practices and policies (Chang and Lin 2007).

[11] See, for example, the research conducted by Kaspersky Lab and B2B International. The study involved 7,993 full-time employees which where asked about policies and responsibilities for corporate IT security. According to this study, just 12% of respondents are fully aware of the IT security policies in the organizations they work for https://www.kaspersky.com/about/press-releases/2017_organizations-it-security-policies.

### 4.5.5 *Look Elsewhere for Inspiration*

In a renewed perspective, it would be useful to take cues from approaches used in other fields which, however, are dealing with security and safety issues. In fact, we need to try to overcome some stereotypical views that, in the end, turn out to be inadequate to handle security problems; on the contrary, it appears more productive to look at positive experiences as a source of inspiration.

From this perspective, Occupational Safety and Health (OSH) shows significant applications and programmes to obtain a safe and healthy work environment. In this area, managing human failure is considered an activity fundamental to prevent accidents; successful experiences in this field can provide some indications for cybersecurity strategy.

The BBS (Behaviour Based Safety) approach is used in industry to prevent accidents and injuries, and can provide examples of good practices. This approach looks at the different aspects of the problem to be handled, hence requiring an adequate assessment of the organization, and not only training. In this sense, we cannot ignore that assessment is one of the most important elements for the building of a CSC, since this is essential to identity the most appropriate measures and initiatives.

Moreover, the focus of BBS approach is on the whole organization and the development of a "total safety culture", where people consider safety as a value (Health and Safety Authority 2013); the meaning is that adopting a proper behaviour ensures their safety but also the safety of others. It is evident that since the protection of information is a priority for modern organizations, a "total CSC" needs employees consider company information (and their personal information) as something precious to protect.

Experiences which combine security and safety issues are another important source of inspiration. We can think, for example, of bank robberies management[12]: even if they appear in decline in comparison with digital crimes, robberies still represent a problem for financial institutions. Here we want to emphasize the combined approach applied to handle this crime by bank organizations. In fact, since the primary goal is to protect the safety and security of personnel and customers, several technical and organizational measures have been implemented over the years. One of the most important measures is to train employees in order to make them aware of the implications of the criminal event for their safety. We are aware that we are dealing with a totally different crime, but the philosophy of an approach based on the combination of safety and security principles can offer some helpful thoughts replicable—with due adaptations—in cybersecurity field. In fact, whether security or safety, human behaviour is the best weapon to respond to threats.

[12]The author has been collaborating for almost fifteen years with the Italian Banking Association (ABI) for the management of physical security issues. In particular, she is the author of a national guide for bank robbery prevention, based on an integrated approach between physical measures and proper employees' behaviour. Moreover, many activities to develop awareness in this area have been developed during these years.

Finally, we must not forget that security and safety are two sides of the same coin: a positive work climate can also have beneficial effects on people's behaviour related to security, as discussed in the following paragraph.

## 4.6 Promoting a Healthy Work Environment

The development of Cybersecurity Culture should not be a separate process from work environment. Like a seed growing in the good soil, CSC needs a receptive environment to flourish effectively. With this concept we refer both to the necessary cooperation between management and employees (ENISA 2018), and to the importance of maintaining a healthy workplace in the broadest sense of the word.

We can assume that if people perceive a positive work environment, they are more likely to carry out their tasks and, probably, they are more willing to adopt secure behaviour.

The question is: how can we expect attentive behaviour in workplace when people are unsatisfied with their job, demotivated or too stressed?

On the other hand, work environment includes different components: technical, human and organizational (Opperman 2002); therefore, several factors affecting employees' performance and behaviour have to be considered.

According to research, a psychologically safe environment, able to include and support employees, can improve their performance (Chandrasekar 2011) and job satisfaction (Raziq 2015). Conversely, job satisfaction has a positive correlation with compliant behaviour (e.g. Kaplan et al. 2009).

Moreover, several studies support the positive relationship between security culture, job satisfaction and compliant security behaviour (Chan et al. 2005; Green and D'Arcy 2010). This means that when people are satisfied with their jobs, they tend to adhere to corporate policies, probably for the reciprocal effect generated by the positive work environment (Settoon et al. 1996).

Some issues related to human nature are sometimes disregarded rather than managed properly. For instance, while conflicts between individuals and groups within organizations can be considered a natural condition related to the different views, their effective management has an important role in ensuring a collaborative workplace. In fact, even if conflicts can be of different types and generate negative—but also positive—effects (Ongori 2009), they can interfere with processes and groups within organization, impeding the achievement of organizational goals.

It is therefore evident that a dysfunctional work environment can also be an obstacle for the handling of security issues. Poor motivation and a lack of cohesion can affect individuals' security behaviour, such as the compliance with the rules. In addition, as discussed in Chap. 2, Sect. 2.2, among the motivations able to induce insiders to illegal actions, there is also individual's dissatisfaction about their working conditions (Kont et al. 2015).

Enhancing employees' motivation is therefore fundamental for effective management and for the development of a cooperative workplace. Even if the financial element is recognized as a strong motivator, it is not the only one, since other factors, like achievement, recognition for achievement, advancement, and the work itself (Herzberg 1987) are still significant.

Currently, organizations are focused on developing strategies to win in highly competitive markets, but few of them give a proper consideration to the fact that human beings are the ones having the real power to define success or decline (Dobre 2013).

Developing a positive and healthy work environment is crucial to promote security awareness initiatives. Passing from the mere fulfilment of laws to a broader and shared approach about security requires a receptive ground, where concepts and policies are understood and spread.

Companies are more than structures, norms and policies: they represent psychological entities where human beings work, create relationships and experience different emotions and perceptions. The term typically used to define this state of emotional health is “organizational climate”, a concept strictly connected to the organizational culture: if organizational culture represents the identity of an organization, climate is what employees perceive regarding their work environment, thus including practices and procedures. In short, we can say that the climate emphasizes the importance of social and cultural components in the workplace.

Widely discussed in the 1960s in the field of organizational behaviour, organizational climate is an important factor for employees’ commitment (Berberoglu 2018). As a multidimensional concept (Litwin and Stringer 1968), it includes both the organisational structure and workers’ perceptions relative to work conditions.

Large companies are used to develop periodical organizational climate surveys to measure work climate and monitor its evolution. These surveys can be compared to the weather forecast, but instead of predicting if it rains or not, they can take the current state and verify the level of acceptance of new changes.

Climate survey is a helpful tool for measuring the effectiveness of awareness outcomes of practices applied in the past; at the same time, it clarifies the level of involvement and engagement with which workers manage the relationship with their organization. In fact, the dimensions investigated in a climate survey generally include:

- organizational dimensions, such as clarity of functions and corporate roles, system of awards and incentives;
- interpersonal dimensions, for example quality of the relationship among workers and between workers and leadership, communication, cooperation;
- individual dimensions, such as sense of belonging, motivation, responsibility, work satisfaction.

Each dimension is significant for building a positive work climate. We know that a sense of belonging is an essential human need, widely emphasized in the popular “Theory of human needs” (Maslow 1970). According to the pyramid model, a sense of belonging is placed between physiological needs and safety, esteem and self-actualization.

Being part of a group or an organization makes people feel accepted and increases their motivation; consequently, they are more likely to be productive and cooperative in working with others. The more they feel a sense of belonging, the greater their commitment to take pride of their organization.

Sense of belonging, for example, is a strategic factor when employees are called to develop careful behaviours to protect what they belong to. According to research (e.g. Cheng et al. 2013; Ifinedo 2014), employees feeling a strong attachment to their organization (identification with company values) tend to comply with security policies. Moreover, sense of belonging can be a trigger for security actions (Pfleger et al. 2014), such as in the case of the protection of symbols representing the organizational brand (e.g. websites, logos).

Unfortunately, the society we live in seems to have lost some basic principles, and people are often considered as numbers or simple gears of the business machine: what is important is that the machine works, no matter what.

Work environment continues to change, and despite the reassurances about the future of work, job insecurity represents a concern for people. Globalization, different market labours, organizational changes, technological applications are producing diverse job expectations. At the same time, demanding requests are posed to workers regarding their skills, competences and performance. In addition, work-related stress and information overload are a menace for people's health (Sect. 2.6).

Companies interested in increasing their business and improving their level of cybersecurity should also invest in their employee's well-being and in a positive organizational climate. Motivated people can make the difference in any organizational security strategy.

## References

Al Hogail, A.: Cultivating and assessing organizational information security culture, an empirical study. Int. J. Secur. Appl. **9**, 163–178 (2015)

Badal, S., Harter, J.K.: Gender diversity, business-unit engagement, and performance. J. Leadersh. Organ. Stud. **21**(4), 354–365 (2014)

Barney, J.B.: Organizational culture: can it be a source of sustained competitive advantage? Acad. Manag. Rev. **11**, 656–665 (1986)

Bear, J.B., Woolley, A.W.: The role of gender in team collaboration and performance. Interdisc Sci Rev. **36**(2), 146–153 (2013)

Beautement, A., Sasse, M.A., Wonham, M.: The compliance budget: managing security behaviour in organisations. ACM Press, New York (2008)

Berberoglu, A.: Impact of organizational climate on organizational commitment and perceived organizational performance: empirical evidence from public hospitals. BMC Health Serv. Res. **18**(1), 399 (2018)

Cameron, K. S., Quinn, R. E.: Diagnosing and Changing Organizational Culture: Based on the Competing Values Framework. Wiley (2011)

Chan, M., Woon, I., Kankanhalli, A.: Perceptions of information security at the workplace: linking information security climate to compliant behavior. J. Inf. Priv. Secur. **1**(3), 18–41 (2005)

Chandrasekar, K.: Workplace environment and its impact on organizational performance in public sector organizations. Alagappa University, Karaikudi, India (2011)

Chang, S.E., Lin, C.-S.: Exploring organizational culture for Information Security Management. Ind. Manag. Data Syst. **107**, 438–458 (2007)

Chen L. C., Farkas D.: An investigation of decision-making and the tradeoffs involving computer security risk. In: Proceedings of the 15th Americas Conference on Information Systems, vol. 610 (2009)

Cheng, L., Li, Y., Li, W., Holm, E., Zhai, Q.: Understanding the violation of IS security policy in organizations: an integrated model based on social control and deterrence theory. Comput. Secur. **39**(Pt B), 447–459 (2013)

Corradini, I., Nardelli, E.: A case study in information security awareness improvement: a short description. In: Proceedings of IMCIC-ICSIT, Florida, USA, pp. 184–186 (2016)

Corradini, I., Nardelli, E.: Building organizational risk culture in cyber security: The role of human factors. In: Ahram T., Nicholson, D. (eds.) Advances in Human Factors in Cybersecurity, AHFE 2018. Advances in Intelligent Systems and Computing, vol 782. Springer, Cham (2019)

Cyber Security Breaches Survey: UK Department for Digital, Culture, Media and Sport, Ipsos MORI. University of Portsmouth, Cyber Security Breaches Survey (2019). https://assets.publishing.service.gov.uk/government/uploads/system/uploads/attachment_data/file/875799/Cyber_Security_Breaches_Survey_2019_-_Main_Report_-_revised.pdf

Da Veiga, A.: Achieving a security culture. In: Vasileiou, I., Furnell, S. (eds) Cybersecurity Education for Awareness and Compliance, Information Science Reference (2019)

Da Veiga, A., Eloff, J.H.P.: An information security governance framework. Inf. Syst. Manag. **249**(4), 361–372 (2007)

Da Veiga, A., Eloff, J.H.P.: A framework and assessment instrument for Information security culture. Comput. Secur. **29**, 196–207 (2010)

Davis, S.M.: Managing Corporate Culture. Ballinger, Cambridge, MA (1984)

Dhillon, G.: Interpreting the Management of Information systems Security. London School of Economics and Political Science, London (1995)

Dobre, O.I.: Employee motivation and organizational performance. Rev. Appl. Socio Econ. Res. **5**(6), 53–60 (2013)

Eftekhari, P., Spaniel, D.: The business value of a diverse infosec team, ICIT (2020). https://icitech.org/wp-content/uploads/2020/01/ICIT-Brief-The-Business-Value-of-a-Diverse-InfoSec-Team.pdf

ENISA: Cyber Security Culture in Organisations: (2018) https://www.enisa.europa.eu/publications/cyber-security-culture-in-organisations

ENISA: Cybersecurity Culture Guidelines: Behavioural Aspects of Cybersecurity: (2019) https://www.enisa.europa.eu/publications/cybersecurity-culture-guidelines-behavioural-aspects-of-cybersecurity

Fitzgerald, T.: Building management commitment through security councils, or security council critical success factors. In Tipton, H. F. (ed) Information Security Management Handbook, pp. 105–121. Auerbach Publications, Hoboken (2007)

Foley, M., Dewey, L., Williamson, S.F., Blackman, D., Creagh, A., Davidson, L., et al.: Women in cyber security literature review (2017)

Goudenogh, W.H.: Componential analysis and the study of meaning. Language **32**, 195–2016 (1956)

Greene, G., D'Arcy, J.: Assessing the impact of security culture and the employee-organization relationship on IS security compliance. In: Proceedings of the Fifth Annual Symposium on Information Assurance. Albany, NY (2010)

Groysberg, B., Lee, J., Price, J., Cheng, J.Y.: The leader's guide to corporate culture. Harvard Bus. Rev. **96**, 44–52 (2018)

Halevi, T., Memon, N., Lewis, J., Kumaraguru, P., Arora S, Dagar, N., et al.: Cultural and psychological factors in cyber-security. In: Proceedings of 18th International Conference on Information Integration and Web-based Applications and Services, vol. 13, pp. 43–56 (2017)

Hatch, M.J.: Organization Theory. Modern, Symbolic, and Postmodern Perspectives. Oxford University Press, New York (1997)

Hayden, L.: People-Centric Security Transforming Your Enterprise Security Culture. MacGrawHill Education, New York (2015)

Health and Safety Authority (HAS): Behaviour Based Safety Guide (2013)

Herzberg, F.: One more time: how do you motivate employees? Harvard Bus. Rev. **65**, 5 (1987)
Hinson, G.: Human Factors in Information Security. IsecT Ltd (2003)
Hofstede, G.: Cultural Consequences: International Differences in Work Values. Sage, Beverly Hills (1984)
Hofstede, G., Hofstede, G. J. & Minkov, M.: Cultures and Organizations: Software of the Mind (Rev. 3rd ed.). New York: McGraw-Hill (2010)
Hu, Q., Dinev, T., Hart, P., Cooke, D.: Managing employee compliance with information security policies: the critical role of top management and organizational culture. Decis. Sci. J. **43**(4), 615–60 (2012)
Huang, K., Pearlson, K.: For what technology can't fix: building a model of organizational cybsersecurity culture. In: Proceedings of the 52nd Hawaii International Conference on System Sciences (2019)
Ifinedo, P.: Information systems security policy compliance: an empirical study of the effects of socialisation, influence, and cognition. Inf. Manag. **51**(1), 69–79 (2014)
Kaplan, S., Bradley, J.C., Luchman, J.N., Haynes, D.: On the role of positive and negative affectivity in job performance: a meta-analytic investigation. J. Appl. Psychol. **94**(1), 162–176 (2009)
Kearney, P. (2010). Security: The Human Factor. IT Governance Publishing, Cambridgeshire (2010)
Kont, M., Pihelgas, M., Wojtkowiak, J., Trinberg, L., Osula, A.M.: Insider Threat Detection Study. NATO CCD COE, Tallinn (2015)
Laycock, A., Petric G., Roer K.: The seven dimensions of security culture. CLTRe (2019)
Lim, J.S., Chang, S., Maynard, S.B., Ahmad, A.: Exploring the Relationship between Organizational Culture and Information Security Culture, 7th Australian Information Security Management Conference, SECAU Security Congress 2009. Perth, Western Australia (2009)
Litwin, G.H., Stringer, R.A.: Motivation and Organizational Climate. Harvard Business School Press, Cambridge, MA (1968)
Lundy, O., Cowling, A.: Strategic Human Resource Strategy. Routledge, London (1996)
Marshak, J.R.: Covert Processes at Work: Managing the Five Hidden Dimensions of Organizational Change. Berrett-Koehler, San Francisco (2006)
Martins, N., Da Veiga, A.: An information security culture model validated with structural equation modelling. In: Proceedings of the Ninth International Symposium on Aspects of Information Security & Assurance, HAISA (2015)
Martins, A., Eloff, J.: Information Security Culture, Security in the Information Society, pp. 203–214, IFIP/SEC 2002. Kluwer Academic Publishers, Boston, MA (2002)
Martin, J., Siehl, C.: Organizational culture and counterculture: an uneasy symbiosis. Organ. Dyn. **12**(2), 52–64 (1983)
Maslow, A.H.: Motivation and personality. 2nd ed. New York: Harper & Row (1970)
Mc Kinsey & Company: Perspectives on transforming cybersecurity (2019). https://www.mckinsey.com/~/media/McKinsey/McKinsey%20Solutions/Cyber%20Solutions/Perspectives%20on%20transforming%20cybersecurity/Transforming%20cybersecurity_March2019.ashx
O'Flaherty, K.: How diversity can help fight cyber-attacks (2018). https://www.information-age.com/how-diversity-can-cyber-123477494/
Okere, I., van Niekerk, J., Carroll, M.: Assessing information security culture: A critical analysis of current approaches. In: Proceedings of IEEE Conference on Information Security for South Africa (ISSA), pp. 1–8, South Africa (2012)
Ongori, H.: Organizational conflict and its effects on organizational performance. Res. J. Bus. Manag. **2**(1), 16–24 (2009)
Oost, D., Chew, E.: Investigating the Concept of Information Security Culture: UTS: School of Management Working Paper: No. 2007/6 (2007)
Opperman, C.S.: Tropical business issues. Partner Price Water House Coopers, International Business Review (2002)
Pfleeger, S.L., Sasse, M.A., Furnham, A.: From weakest link to security hero: transforming staff security behavior". J. Homel. Secur. Emerg. Manage. **11**(4), 489–510 (2014)

Quinn, R.E., Rohrbaugh, J.: A spatial model of effectiveness criteria: towards a competing values approach to organizational analysis. Manag. Sci. **29**(3), 363–377 (1983)

Raziq, A., Maulabakhsh, R.: Impact of working environment on job satisfaction. Procedia Econ. Financ. **23**, 717–725 (2015)

Robbins, S.P.: Organizational Behavior. Prentice Hall, Upper Saddle (2001)

Rock, D., Grant, H.: Why diverse teams are smarter. Harvard Bus. Rev. (2016). https://hbr.org/2016/11/why-diverse-teams-are-smarter

Schein, E.H.: Organisational Culture and Leadership. Jossey Bass, San Francisco (1985)

Schein, E.H.: The Corporate Culture Survival Guide: Sense and Nonsense about Culture Change. Jossey-Bass Inc., San Francisco (1999)

Schlienger, T.: Informationssicherheitskultur in Theorie und Praxis: Analyse und Förderung soziokultureller Faktoren der Informationssicherheit in Organisationen. Fribourg, iimt University Press (2006)

Schlienger, T., Teufel, S.: Information security culture—from analysis to change. South Afr. Comput. J. **31**, 46–52 (2003)

Schlienger T., Teufel S.: Tool supported management of information security culture. In: Sasaki R., Qing S., Okamoto E., Yoshiura H. (eds) Security and Privacy in the Age of Ubiquitous Computing, pp 65–77, SEC 2005. IFIP Advances in Information and Communication Technology, vol 181. Springer, Boston, MA (2005)

Schneier, B.: Secrets and Lies. Wiley, New York (2000)

Schultz, E.: The human factor in security. Comput. Secur. 24, 425–426 (2005)

Settoon, R.P., Bennett, N., Liden, R.C.: Social exchange in organizations: perceived organizational support, leader-member exchange, and employee reciprocity. J. Appl. Psychol. **81**(3), 219–227 (1996)

Siehl, C., Martin, J.: The role of symbolic management: How can managers effectively transmit organizational culture? In Hunt, J., Hosking, D., Schriesheim, C., Stewart, R. (eds.) Leaders and Managers: International Perspectives on Managerial Behavior and Leadership. Pergamon, New York, 227–269. (1984)

Siponen, M.T.: A conceptual foundation for organizational information security awareness. Inf. Manag. Comput. Secur. **8**(1), 31 (2000)

Siponen, M.T.: Five dimensions of information security awareness. Comput. Soc. **31**(2), 24–29 (2001)

Tyler, T.R.: Promoting employee policy adherence and rule following in work settings: the value of self-regulatory approaches. 70 Brooklyn Law Rev. 1287–1312 (2005)

Van Niekerk, J., Von Solms, R.: Information security culture: a management perspective. Comput. Secur. **29**, 476–486 (2010)

Wilson, M., Hash, J.: Building an Information Technology Security Awareness and Training Program, National Institute of Standards and Technology (NIST), Special Publication 800-50, Computer Security (October 2003)

# Chapter 5
# Communication is Not an Option

**Abstract** In a psychosocial approach to cybersecurity, we cannot neglect the role played by communication, since it is the key of any security strategy and, generally speaking, it is fundamental in any environment where human beings interact. In this chapter, we discuss the importance of both internal communication, in order to disseminate contents related to cybersecurity, and crisis communication management, when crisis events such as data breaches occur. Firstly, communication supports the building of Cybersecurity Culture through the sharing of common values and beliefs among workers. For this goal, it is important to use clear and understandable communication, since it is not obvious that all employees know technical language. Moreover, for effective communication, cognitive and emotional aspects have to be taken into consideration. For example, since communication about risks can evocate adverse effects, messages need to balance positive and negative elements, in order to put people in the condition to feel able to handle threats; otherwise the real risk is the adoption of careless or passive behaviour. Finally, an effective communication strategy for managing crisis events is essential to preserve company's reputation. In an always-on world the topic of reputation is vital for everyone, individuals and organizations, since it is more and more dependent on people's interactions in the digital world. Data breaches can produce serious impacts for future business: compromising the relationship with their stakeholders, organizations put at risk the opportunities of new trade, and therefore, they should pay great attention to customers data protection.

## 5.1 Communication Skills in Cybersecurity

Communication is the key of any security strategy and, generally speaking, it is fundamental in any environment where human beings interact. For organizations, communication plays a strategic role, both internally and externally, having the power to ensure long-term success.

© Springer Nature Switzerland AG 2020

I. Corradini, *Building a Cybersecurity Culture in Organizations*,
Studies in Systems, Decision and Control 284,
https://doi.org/10.1007/978-3-030-43999-6_5

Here, we first discuss the importance of appropriate communication skills in managing cybersecurity, then we extend our analysis to the role of communication in managing crisis events and protecting company reputation when security incidents occur, as in the case of data breach (Sect. 5.3).

The starting point is that poor and unclear communication can be an obstacle to effective cybersecurity. One issue regards the limited communication between IT managers responsible for cybersecurity and other leaders of the same organization. The pressure to which they are exposed to, along with the weight of responsibility and the challenge to prioritise emerging threats, leads IT managers to feel isolated from others.[1]

Perhaps, precisely for their function, cybersecurity teams are often perceived as something unique and separated from the rest of the organization; to improve communication, some barriers have to be broken down, because to handle a problem—and cyberthreats are a problem for companies—it is necessary, above all, to discuss it. Moreover, since cybersecurity involves all the people working in the organization, sharing communication with all the other internal functions is strongly needed.

Looking at the future of cybersecurity workforce, we should go beyond technical and engineering skills (Dawson and Thompson 2018), and consider the importance of combining them with communication and social abilities. There is little about the skills necessary for successful cybersecurity advocates (Haney and Lutters 2017), but what is clear is that other non-technical skills are required. Many are focused on technical skills, forgetting that an organization is made up of individuals with different level of knowledge and needs. Cybersecurity requires adequate communication skills: technical concepts need to be described and explained in simple words, if everybody has to be able to understand what happens in the cyberworld. Unfortunately, there is not yet an adequate sensitivity to this issue.

On the other hand, communication skills are abilities that everybody uses in providing and receiving different kinds of information: they are part of the so-called transversal skills for doing the job, i.e. team working, problem solving, planning, and so on. Their main characteristics is their high transferability across the different job sectors.

In spite of the increasing digitalization of economy and the changing structure of employment, we cannot exclude the relevance of communication skills for improving people's participation in modern society. Moreover, in relation to digital skills, which are required in all types of jobs (European Commission 2017),[2] we observe that these include, among others, being able to communicate via email or social media and to protect personal information online.

[1]See the research carried out by Opinium and commissioned by Trend Micro (2019). The online survey involved 1,125 IT decision-makers responsible for cybersecurity across the UK, US, Germany, Spain, Italy, Sweden, Finland, France, Netherlands, Poland, Belgium and Czech Republic. https://newsroom.trendmicro.com/press-release/commercial/trend-micro-finds-one-third-cybersecurity-staff-feel-isolated-business.

[2]https://ec.europa.eu/digital-single-market/en/news/new-report-shows-digital-skills-are-required-all-types-jobs.

Consequently, if raising awareness on digital technologies is a priority for living in the digital market, communication is the main means through which it is possible to achieve the expected awareness. Moreover, we must not forget that communication is the basis of effective security awareness campaigns, as well as of effective cybersecurity training programmes.

Among the different communication skills, the most important ones are:

- *Active listening*, which goes beyond passively hearing, involving all senses in an active process and including verbal and not-verbal signals of listening. We know the importance of body language and of not-verbal cues in the communication process; at the same time, active listening provides recognition of other individual's values and needs. In addition, this kind of skill is fundamental for improving team communication.
- *Ability of adopting communication style to different audiences and situations.* Despite the different styles, assertive communication is considered the most effective, compared to others (passive, aggressive and passive-aggressive). Using assertive communication requires understanding how others communicate: in this way messages can reach them and be easily taken on board.
- *Clarity of communication*, understandable to the audience. One thing is to talk to a colleague in our competence area, another is interacting with other functions and roles in the organization. For this purpose, it is fundamental to identify what and how to communicate; for example, using technical terms in messages addressed to people without the appropriate background, can cause communication to fail. In fact, people could be not interested in the subject or not ready to receive the information.
- *Empathy*, one of the most important skills for developing effective communication. In short, empathy means recognizing emotions in others and considering their different perspectives; therefore, it promotes the building of trust relationship. Demonstrating empathy does not mean to agree completely with others, but understanding what they think, feel and experience. We know that empathy takes time and effort, and that it also depends on our personality traits; however, it is a key of success for any relationship. The use of empathy in workplace, for example, help employees to feel appreciated and supported and to be encouraged in their performance.

Some of the aspects discussed above appear particularly significant to make communication in cybersecurity effective. It is adequate when it is clear and understandable to the audience; in this sense, information and messages on cybersecurity should be specifically tailored for different targets to maximize the impact of them.

We cannot expect that all employees have the same level of comprehension when adopting technical language; we therefore should avoid using obscure terminology, insider jargon, or complicated nomenclature, while we should strive to ensure the consistency of messages.

When people acquire adequate information and are familiar with it, they are more likely to develop a favourable attitude towards the content they have received. Familiarity with terminology can help people to recognize certain risks and to develop more sensitivity towards them. The cognitive fluency, i.e. the ease with which information is processed, is typically associated to positive feelings; in fact, fluency is a cue to familiarity (Winkielman et al. 2003) referring to something already encountered before. According to research, feeling of familiarity affects persuasive processing (Garcia-Marques and Mackie 2001).

We know how important risk communication is within organizations, since this can promote a change in individual behaviour and define how the risk will be handled (Infanti et al. 2013; Lundgren and McMakin 2013). From this point of view, risk communication is not just an information activity, but it represents an interaction based on exchange of experiences between individuals.

In addition, this kind of communication is characterized by the involvement of psychological elements connected to the concept of risk and to human nature. In fact, risk calls on elements of uncertainty, while people show an innate need to feel safe.

The final goal of communicating risks, including cybersecurity risks, is therefore not to frighten or even reassure individuals, but induce them to act responsibly and properly. This is why communication has to be effectively designed and implemented, relying on specific methods and tools (Chap. 6).

For effective communication we should consider the importance of cultural and individual differences, and the fact that messages are more effective when they match the recipient's cognitive and motivational characteristics, as well as their risk perception.

Moreover, the process should not be limited to one-way communication, but provide opportunities to share experiences; too often digital communication is preferred over the traditional one, while a combination of different methods and tools is recommended.

Finally, we cannot ignore the importance of emotional elements on individual's responsiveness, especially when media communication emphasises the negative effects of cyberthreats and data breaches (Sect. 5.2).

## 5.2 Beware of the Desensitization Effect

In the era of cyber-attacks and data breaches, the dominant message is that we cannot stop them and therefore we must get used to accepting these events, as part of our everyday lives. Daily and alarming news has contributed to reinforce the belief that privacy and security threats are inescapable.[3]

It might be the price to be paid for the increasing digitalization of our society. However, it does not mean that we have to accept what happens without doing anything. Actually, people are interested in protecting their data: it seems that when a data breach occurs, they prefer companies ensure them that all measures have been

[3]Think, for example, to Facebook-Cambridge Analytica scandal in early 2018. It was revealed that about 87 million user's personal data of Facebook profiles were harvested by Cambridge Analytica and improperly used for political goal, without the consensus of those involved.

taken to avoid another breach rather than receiving a financial compensation for the inconvenience (Ablon et al. 2016).

From a psychological view, it is usual that people feel powerless when certain events occur; even though we proceed with all due caution, the problem could be created by someone else, without our intervention. It seems like the only possible thing we can do is to be prepared to manage the consequences.

In a sense, information about sensational incidents and effects has the power to promote awareness of the cyber-risks we are exposed to. Persuasive messages often emphasise the harm deriving from the non-adoption of recommendations, pointing at individual's fear (Tannenbaum et al. 2015). Looking at the contents of cybersecurity awareness campaigns, in fact, we observe that they are essentially focused on damages connected to users who do not follow right practices (Bada and Sasse 2014). There is, however, another side of the coin. If people perceive that there is no way out of these risks escaping from their control, they will probably end up accepting the situation passively, and this resignation can become the major risk to be handled.

People might wonder why putting so much effort into something which they cannot control. In a nutshell, a sort of a "desensitization" effect: if data leaks are normal and represent an everyday occurrence,[4] individuals can lose interest for cybersecurity issues and decrease attention to secure behaviour.

Such an effect, able to influence people's reaction, is a well-known tendency in psychology. We can think of terrorist attacks: even if they have a strong emotional impact, their frequency decreases the novelty effect, and reduces people's emotional response and interest. We cannot compare this criminal phenomenon to security and privacy problems we are discussing here, but human beings have the tendency to repeat certain reaction patterns.

Hence, given the limited control on one's own private data and the news about these negative events, people could be induced to think that efforts required to manage this situation are beyond their capabilities.

In literature there is not a full agreement about the fact that an appeal to emotion-based arguments can persuade people effectively. On one hand, some authors believe that fear appeal can have a persuasive effect (e.g., Xu et al. 2015); on the other hand, others say that this type of approach can even be counterproductive (e.g., Ruiter et al. 2014).

It is evident that people's reactions are varied and that a person can have different approaches to the same danger. Perhaps, the best way to handle risk communication is to balance the effect of "fear" and the ability of "coping" with it. Messages on cybersecurity risks are likely to be more effective (and persuasive) when they have the ability to induce people to perceive the availability of possible solutions (e.g. Ruiter et al. 2001).

Talking about communication strategies, the reinforcement of positive terms, replacing negative, can encourage constructive behaviour. For example, we are used

[4]https://edgy.app/data-leak-normal-now.

to marketing human beings as the weakest link of security. This is a common stereotype, which can be an obstacle to employees engagement. In fact, the continued attribution of a "negative label" (weakest link in the present case) to individuals can lead them to make errors, since they feel that they are perceived as the problem and not as part of the security solution. Moreover, the risk is that they behave according to the negative expectation.[5]

This negative stereotype can be turned into a positive one: replacing the adjective "weakest" with "strongest" or "reliable" might reinforce the active role of employees in managing security. When employees feel that they are appreciated, they are likely to enhance their sense of belonging to the organization, ensuring their participation in company initiatives.

## 5.3 Protecting Data and Reputation

Over the last few years the importance of reputation has grown, especially because of the role played by social media platforms.

Looking at the available definitions, reputation regarding firms (i.e., corporate reputation) can be interpreted under several perspectives, according to six different disciplines (Fombrun and van Riel 1997): economics, marketing, sociology, organizational behaviour, accountancy and strategy. However, the distinctive element is how stakeholders, in the broadest sense of the word, perceive and judge organization activities; reputation is therefore a collective measure based on company trustworthiness (Jøsang 2007) and, given the nature of different stakeholders, an organization has more than a single reputation (Chun 2005).

A strong reputation produces many advantages for companies (e.g. Fombrun 1996; Feldman et al. 2014), such as greater visibility, protection of their values, the possibility of differentiating them from their competitors, and the improvement of their capacity to retain qualified personnel.

In a global and uncertain market, firms must improve their ability to build and maintain customers' confidence.

Since cybersecurity is a must and customers trust is strategic for business, it is evident the connection between cyberthreats and reputation: companies which cannot protect consumers data might compromise their trust, generating a risk for future business.

---

[5]The Labelling Theory (e.g. Becker 1973) is a sociological approach developed in the 1960s in order to understand deviant behaviour. Among factors influencing criminal behaviour, there is also the way by which people are labelled. This theory is associated with the concept of self-fulfilling prophecy (Rosenthal and Jacobson), negatively labelled individuals are more likely to develop negative behaviour. Even though this theory received several critics, it is however interesting to observe how the repetitive use of negative (and also positive) terms can affect individual self-identity.

Cybersecurity risks are considered an important operational challenge to be handled by global executives,[6] and data security is one of the most relevant key macro-trends for the reputational landscape (Reputation Leader study 2019).[7]

Data breaches, for example, produce both economic damages and negative effects on brand reputation to the point where data privacy and security are perceived by leaders as a priority to manage (Reputation Institute, 2020).

Everyone knows the notable quote by Warren Buffett according to which "it takes 20 years to build a reputation and five minutes to ruin it."[8] This is especially true in the era of social media. Therefore, to achieve and keep a positive reputation, companies have to consider the relationship between this intangible asset and all risks, cyberthreats included.

It is plausible that the increasing attention to reputation is partially associated with the current regulations, in particular with the EU's General Data Protection Regulation (GDPR). This law, adopted in May 2018, has imposed some restrictive requirements for individual data privacy,[9] making organizations accountable; data protection principles have to be necessarily considered in the design of new products or activities (approaches "by design" and "by default"). From this point of view, organizations are encouraged to implement technical and organizational measures at the earliest stages of the design of the processing operations, in order to safeguard privacy and data protection principles from the start (data protection "by design"). Organizations should ensure that personal data is processed with the highest privacy protection in order to avoid that—by default—personal data is accessible to an indefinite number of people (data protection "by default").[10]

In addition, for certain types of personal data breach, posing a risk to the rights and freedoms of individuals, organizations must report them to the relevant supervisory

---

[6]See, for example, the C-Suite 2018, the Annual Survey of Global Business Executives, conducted by Global Business Policy Council in partnership with ATKearney. This report is based on results from a survey including four sections. The first asks executives to assess the likelihood of a variety of discrete potential global developments. The second focuses on the external environment (e.g. economic, political, social) affecting businesses. The third concentrates on business operations, that is firm's internal processes, people, and systems. In the fourth section a topical issue is chosen; for this report the section is based on cybersecurity strategy, since this topic has consistently ranked as a top challenge among executives in past iterations of this survey. From the study, relating to this specific issue, it emerges that cybersecurity tops the list of challenges for business operations. https://www.kearney.com/web/global-business-policy-council/article?/a/2018-views-from-the-c-suite.

[7]The Reputation Leaders study by Reputation Institute (2019) is based on the participation of 179 Global Communication executives. Cyber-attacks and data privacy are the second top Macro-Trends. Among others we can find fake news, female empowerment, CEO activism, and so on.

[8]https://www.forbes.com/pictures/eede45imgh/it-takes-20-years-to-bui/#252b4f98ec35.

[9]According to GDPR, personal data refers to any information that relates to an identified or identifiable, living individual, for example: name, address and phone number, location, health records, income, banking information, cultural preferences, and more https://ec.europa.eu/commission/sites/beta-political/files/data-protection-factsheet-sme-obligations_en.pdf.

[10]https://ec.europa.eu/info/law/law-topic/data-protection/reform/rules-business-and-organisations/obligations/what-does-data-protection-design-and-default-mean_en.

authority, no later than 72 hours after having become aware of the fact (article 33, GDPR).

We know that when a company admits to have suffered from a data breach, this generates doubts about its capability of protecting customer data. Actually, from a reputational viewpoint, there can be a twofold problem: even if with different responsibilities, both customers and company are victims of the attack, and both parties suffer from the consequences. It is therefore evident that, besides offering products and services, an organization should guarantee the best protection for its customers and their data.

Considering the challenges of the global market and the current regulations, it is clear that the achievement of customers confidence also passes through the ability of ensuring data security and privacy. Whether physical or cyber, security is one of the key drivers of reputational risk (Deloitte 2014), and each organization should consider the impact from unpredictable cyberthreats.

Studies on costs related to data breaches reveal that these generate at least four types of additional costs (Ponemon Institute 2018):

- detection and escalation (e.g. forensic and investigative activities, crisis team management);
- notification (e.g. communications with regulators, engagement with outside experts);
- post data breaches response (e.g. help desk activities, legal expenditures);
- lost business cost (e.g. cost of business disruption, reputation losses and diminished goodwill).

Some examples that have occurred may help us to understand the relationship between cybersecurity and reputation.

In April 2011 Sony was struck by a data breach where personal details of approximately 77 million accounts were compromised, and the company's servers were out of service for about a week. At first, Sony acknowledged on the official PlayStation Blog that certain functions of the Play Station Network were down. Later, the company announced that an external intrusion had affected the PlayStation Network and Qriocity services.[11]

The management of the breach produced an impact on Sony's reputation; according to the Global RepTrack 100 (2012) by Reputation Institute,[12] after the cyberattack Sony slipped down in the list of the top 100 companies with the best reputation in the world, from the second position in 2010 to the sixth in 2011. The change in position is not in itself of large entity, but it translates into economic damages and loss of consumer confidence.

Another interesting case regards the massive security breach that struck in 2015 Ashley Madison, a famous dating website for married people: a group of hackers,

[11] https://en.wikipedia.org/wiki/2011_PlayStation_Network_outage.

[12] A global company specialized on reputation measurement and management. Its model for measurement company's reputation (RepTrack Model) is a standardized framework internationally recognised.

called Hacking Team, hacked the site, exposing about 36 million user accounts all over the world, including names, banking data, credit card transactions, and other sensitive personal information, like secret sexual fantasies.[13] From an economic point of view, the owner of the Ashley Madison website was ordered to pay $11.2 million.[14] In addition, the company who owned and operated Ashley Madison—Avid Life Media—changed its name, now Ruby Corp, probably to avoid the association between the website and the data breach news. Finally, other significant consequences seem to be associated to this event, given that, according to the available information, two people committed suicide after their names were made public.[15]

Considering what discussed above, it is evident that effective cybersecurity contributes to develop and maintain company reputation, and it could represent an important indicator for its measurement (Corradini and Nardelli 2020).

Finally, for effective reputation management companies should not forget the strategic role of communication, given that omissions and mistakes in this area can cause more damages than the cyber-attack itself. Crisis communication is therefore the other important issue we focus on this chapter.

## 5.4 Crisis Communication

Communication can have a strategic role in crisis management, especially when cyber-attacks occur and end up in the public domain. As they usually say, sooner or later, any organization will have to manage a crisis event. What makes the difference is the ability to act quickly and decisively, so as to reduce negative impacts.

When a company crisis breaks out, whatever its nature might be, adequate communication should be enabled, since crises represent a threat to the organizational reputation (Coombs 2007).

Crisis communication is an integral part of crisis management, a proactive process whose goal is to prevent or reduce the damage associated to the crisis (Zamoum and Gorpe 2018).

With the advent of social media, the paradigm of communication has been shifted; in handling a crisis event, the role of communication is extremely delicate, because when it is mismanaged, the risk is to increase the gravity of the crisis itself. As discussed (Chap. 1, Sect. 1.5), social media has the power of intensifying negative (and positive) emotions. Inadequate responses or errors caused by an excess of impulsivity are more likely to produce uncontrollable consequences. Virality and the need of responding to users in real time requires a continuous attention in order to avoid

[13] Hackers justified their actions on the grounds that Ashley Madison lied to users and didn't protect their data properly.

[14] https://www.reuters.com/article/us-ashleymadison-settlement-idUSKBN19Z2F0.

[15] https://www.bbc.com/news/technology-34044506.

irreparable harm. We are aware that media are typically interested in negative news, because of emotional implications and of the asymmetric responses: there is a human tendency to focus its attention on negative contents, since, in evaluating cost-benefit, the costs surpasses the benefits of positive news (Soroka 2006).

The appropriate use of social media is strategic for an effective handling of communication during a crisis. Regardless of the type of communication chosen by the organization, we think that specific actions have to be avoided: for example, denying, accusing someone else, declining all responsibilities about the incident, do not usually produce effective results. At the same time, providing wrong or discordant information reveals how poor and unclear the communication strategy is.

Communication crosses all phases of crisis management and supports them. Different models are available to handle crises, and the same procedures are not always shared by experts.

On the basis of the wide available literature (e.g. Fink 1986; Pearson and Mitroff 1993; Seeger et al. 2003) we can summarize the main phases of crisis management:

- preparation
- response
- recovery

The importance of the preparation phase is often appreciated only after an event has occurred. It is well known, in theory, that prevention is always the best answer to face threats; unfortunately, in practice, this often is not carried out. Especially in cybersecurity, the tendency is to focus above all on how organizations can continue to operate after a cyber-attack. It is true that certain cyberthreats are inevitable, but investing resources on prevention is never a waste.

Generally speaking, we are getting used to preferring emergency management over prevention, without considering the difference in terms of costs and visibility. It is not a simple task to calculate saved costs deriving from effective prevention; on the contrary, we can easily identify the investment for managing a crisis event. In this case, then, we should include some aspects not immediately quantifiable, like medium and long-term impacts on reputation (Corradini and Nardelli 2015).

However, investing in the preparatory phase means being ready to tackle a crisis and limit as far as possible its impact. In this stage a set of strategic activities is developed, such as: creation of the crisis management team; evaluation of the vulnerable areas of the organization; monitoring of signals; definition of crisis communication plan.

The more careful and planned the preparation phase is, the more effective responses to crisis will be.

During the response phase, the identified activities are implemented, with the aim of ensuring company's business continuity. The effectiveness of this stage is strictly related to the accuracy of the preparatory phase; in addition, it is important to bear in mind that other inconveniences and unforeseen events might occur, requiring further efforts.

Finally, in the recovery phase, activities are carried out to restore the initial condition. This stage has to be interpreted as a growth experience for the organization; indeed, management should look at the perspective that what happened could repeat again, in other forms and other ways. In this sense, experience can be transformed into an advantage, contributing to the building of a "toolbox". This is what we call resilience, an important issue, which can be analysed from individual and organizational perspectives. Individual resilience refers to the ability to recover from adverse situations and to become stronger (Walsh 1998); in an organizational context, resilience is the capacity of organizations to continue to be efficient while handling crises or stressful situations to adapt to the demands of the environment (McDonald 2006).

Communication is an essential component of all the three phases above discussed, since it contributes to contain the negative effects associated to the crisis and—at the same time—helps in satisfying the need of keeping stakeholders informed. In this sense, communication strategies should consider both internal and external reactions: worrying only about external stakeholders, and not about what happens internally, means having a partial view of the problem. On the contrary, a crisis invests also employees, whatever their role. They have therefore to be sensitized to appropriately use social media in relation to a crisis involving their organization. They have to be aware that their behaviour may support the crisis management communication strategy.

Moreover, since employees' perception about their organization can be modified because of rumours related to the crisis, communication addressed to them has to be properly handled during the crisis event (Johansen et al. 2012).

Besides all the aspects discussed until now, another important issue is the adoption of clear and understandable communication. In particular, considering the importance of skills discussed in the Sect. 5.1, communication should be (Coombs and Holladay 2010; Reich et al. 2011):

- complete: information about the facts must be exhaustive and make clear what happened;
- concise: few words describing the facts and appealing the audience;
- incisive: messages must be convincing relating to the description of the events and to measures which have to be activated.
- concrete: messages have to be clear, considering the different targets
- timeliness: replies must be given quickly.

Providing the best solution to a crisis also means considering customers' concerns. After a significant data breach, communication should not be limited to invite people to change their password, perhaps through a press release. Instead, communicating emotions, like regret about the event, may reduce people anger and promote the acceptance of the organizational message (van der Meer and Verhoeven 2014). The importance of demonstrating empathy in crisis communication is connected to the attention paid by the organization towards its customers.

In conclusion, to be effective, crisis management should integrate processes, activities, decision-making mechanisms and communication plans into an organic framework. On the one hand, the goal is to preserve business continuity and, on the other hand, to reduce economic, social and reputational impacts deriving from the incidents occurred.

A timely and transparent communication, addressed to both internal and external stakeholders, is crucial to promote a global positive perception of the company ability in handling adverse events effectively.

## References

Ablon, L., Heaton, P., Lavery, D., et al.: Consumer Attitudes Toward Data Breach Notifications and Loss of Personal Information. The RAND Corporation, Santa Monica, CA (2016)

Bada, M., Sasse, A.: Cyber Security Awareness Campaigns: Why Do They Fail to Change Behaviour? Global Cyber Security Capacity Centre. University of Oxford, Oxford, UK (2014)

Becker, H.S.: Outsiders Studies in the Sociology of Deviance. New York: The Free Press (1973)

Coombs, W.T.: Protecting organization reputations during a crisis: the development and application of situational crisis communication theory. Corp. Reput. Rev. **10**(3), 163–176 (2007)

Coombs, W.T., Holladay, S.J.: The Handbook of Crisis Communication, Wiley-Blackwell, John Wiley & Sons, Ltd, Publication (2010)

Corradini, I., Nardelli, E.: La reputazione aziendale. Aspetti sociali, di misurazione e di gestione, Franco Angeli (2015)

Corradini, I., Nardelli, E.: Is data protection a relevant indicator for measuring corporate reputation? Submitted to: 6th International Conference on Human Factors in Cybersecurity, AHFE, San Diego, California, USA, 16–20 July 2020

Chun, R.: Corporate reputation: meaning and measurement. Int. J. Manag. Rev. **7**(2), 91–109 (2005)

Dawson, J., Thomson, R.: The future cybersecurity workforce: going beyond technical skills for successful cyber performance. Front. Psychol. **9**, 744 (2018)

De Loitte Report: Global survey on reputation risk (2014) https://www2.deloitte.com/content/dam/Deloitte/pl/Documents/Reports/pl_Reputation_Risk_survey_EN.pdf

European Commission: New report shows digital skills are required in all types of jobs. (2017). https://ec.europa.eu/digital-single-market/en/news/new-report-shows-digital-skills-are-required-all-types-jobs

European Commission: The GDPR: new opportunities, new obligations. What every business needs to know about the EU's General Data Protection Regulation (2018). https://ec.europa.eu/commission/sites/beta-political/files/data-protection-factsheet-sme-obligations_en.pdf

Feldman P.M., Bahamonde R.A., Bellido I.V.: A new approach for measuring corporate reputation, RAE. Revista de Administracao de Empresas **54**(1), 53–66 (2014)

Fink, S.: Crisis Management: Planning for the Inevitable. AMACOM, New York (1986)

Fombrun, C.: Reputation: Realizing Value from the Corporate Image. Harvard Business School Press, Boston, MA (1996)

Fombrun, C.J., van Riel, C.B.M.: The reputational landscape. Corp. Reput. Rev. **1**(1/2), 6–13 (1997)

Garcia-Marques, T., Mackie, D.M.: The feeling of familiarity as a regulator of persuasive processing. Soc. Cogn. **18**, 9–34 (2001)

Haney, J.M., Lutters, W.G.: Skills and characteristics of successful cybersecurity advocates. In: Third Workshop on Security Information Workers (2017)

Infanti, J., Sixsmith, J., Barry, M.M., Núñez-Córdoba, J., Oroviogoicoechea-Ortega, C., Guillén-Grima, F.: A Literature Review on Effective Risk Communication for the Prevention and Control of Communicable Diseases in Europe. Stockholm, ECDC (2013)

Johansen, W., Aggerholm, H.K., Frandsen, F.: Entering new territory: a study of internal crisis management and crisis communication in organizations. Public Relat. Rev. **38**(2), 270–279 (2012)

Jøsang, A.: Trust and reputation systems. In: Foundations Security Analysis Design IV, FOSAD 2006/2007-Tutorial Lectures, (Bertinoro, Italy), Springer LNCS 4677 (2007)

Lundgren, R.E., McMakin, A.H.: Risk Communication: A Handbook for Communicating Environmental, Safety, and Health Risks. Wiley, Hoboken, NJ (2013)

McDonald, N.: Organizational resilience and industrial risk. In: Hollnagel, E., Woods, D.D., Leveson, N. (eds) Resilience Engineering, Concepts and Precepts, pp. 155–180. Ashgate Publishing Limited, Hampshire (2006). ISBN 0-7546-4641-6

Pearson, C. M., Mitroff, I.I.: From crisis prone to crisis prepared: a framework for crisis management. Acad. Manag. Rev. **7**(1), 48–59 (1993)

Ponemon Institute: Cost of a data breach study: global overview (2018). https://www.ibm.com/downloads/cas/861MNWN2

Reich, Z., Bentman, M., Jackman, O.: A crisis communication guide for public organisations. In: Vos, M., Lund, R., Reich, Z., Harro-Loit, H. (eds) Developing a Crisis Communication Scorecard. Outcomes of an International Research Project 2008–2011, Jyväskylä Studies in Humanities 152, pp. 265–324. Jyväskylä University Press, Jyväskylä (2011)

Reputation Institute: Global RepTrack 100. The World's Most Reputable Companies (2012)

Reputation Institute: What's on the mind of the CCO when it comes to corporate brand reputation? Reputation Leaders study 2018 (February, 2019)

Reuters: Ashley Madison parent in 11, 2 million settlement over data breach (2015). https://www.reuters.com/article/us-ashleymadison-settlement-idUSKBN19Z2F0. Accessed 14 July 2017

Ruiter, R.A.C., Kok, G., Verplanken, B., Brug, J.: Evoked fear and effect of appeals on attitudes to performing breast self-examination: an information-processing perspective. Health Educ. Res. **16**, 307–319 (2001)

Ruiter, R.A., Kessels, L.T., Peters, G.J.Y., Kok, G.: Sixty years of fear appeal research: current state of the evidence. Int. J. Psychol. **49**, 63–70 (2014)

Seeger, M.W., Sellnow, T.L., Ulmer, R.R.: Communication and Organizational Crisis. Quorum Press, Westport, CT (2003)

Soroka, S.N.: Good news and bad news: asymmetric responses to economic information. J. Polit. **68**(2), 372–385 (2006)

Tannenbaum, M.B., Hepler, J., Zimmerman, R.S., Saul, L., Jacobs, S., Wilson, K., Albarracín, D.: Appealing to fear: a meta-analysis of fear appeal effectiveness and theories. Psychol. Bull. **141**(6), 1178–1204 (2015)

van der Meer, T.G.L.A., Verhoeven, J.W.M.: Emotional crisis communication. Public Relat Rev. **40**, 526–536 (2014)

Walsh, F.: Strengthening Family Resilience. Guilford Press, New York (1998)

Winkielman, P., Schwarz, N., Fazendeiro, T.A., Reber, R.: The hedonic marking of processing fluency: implications for evaluative judgment. In: Musch, J., Klauer, K.C. (eds.) The Psychology of Evaluation: Affective Processes in Cognition and Emotion, pp. 189–217. Lawrence Erlbaum, Mahwah, NJ (2003)

Xu, X., Alexander Jr., R.L., Simpson, S.A., Goates, S., Nonnemaker, J.M., Davis, K.C., McAfee, T.: A cost-effectiveness analysis of the first federally funded antismoking campaign. Am. J. Prev. Med. **48**, 318–325 (2015)

Zamoum, K., Gorpe, T.S.: Crisis management: a historical and conceptual approach for a better understanding of today's crises (2018) (interchopen.com)

# Chapter 6
# Developing Cybersecurity Awareness

**Abstract** For an effective Cybersecurity Culture, it is fundamental to develop adequate cybersecurity awareness programmes. Awareness, training and education are the three areas which define a learning process. Understanding their meaning and their mutual relationship is the key for organizations to identify the appropriate tools and methods to induce people to behave responsibly. With awareness we refer to having knowledge of a certain situation and behaving consequently; training is an active and a more or less formal process to teach skills; education is the process of providing integrated knowledge and skills and the means to extend them. Talking of the building of a Cybersecurity Culture, our commitment cannot be limited to the use of materials like videos or posters; these are important in the context of awareness initiatives, but they must be part of a more complex process, whose goal is to induce people to change their behaviour. The success of this process depends on the quality of used tools and contents and on employees' motivation, which has to be stimulated so as to ensure their active participation. Changing insecure behaviour and fostering a responsive mindset is a challenge that cannot be achieved in a short period; this also requires the knowledge of human nature and its mechanisms. In this sense, common habits in security deriving from the physical world can be a source of inspiration for the development of cybersecurity awareness programmes. Finally, some recommendations are provided in order to plan cybersecurity initiatives and to avoid their failure.

## 6.1 The Concept of Awareness

Over the last few years many international initiatives and campaigns have underlined the importance of increasing people's awareness to face cyber-risks. For instance, every year in October several countries develop specific events to raise awareness about cybersecurity among citizens, like the "National Cybersecurity Awareness Month" (NCSAM, United States), and the "European Cybersecurity Month" (ECSM, Europe).

On the other hand, important international regulations such as GDPR defines legal rules to protect personal data and require companies to be compliant, by taking all

© Springer Nature Switzerland AG 2020

I. Corradini, *Building a Cybersecurity Culture in Organizations*,
Studies in Systems, Decision and Control 284,
https://doi.org/10.1007/978-3-030-43999-6_6

necessary measures, technical and organisational, to ensure a high level of information protection. In addition, a European Union (EU) provision (NIS Directive 2016)[1] provides indications to EU countries to strengthen the overall level of cybersecurity and reinforce their cybersecurity approaches at international level. Among the actions considered in the NIS Directive, there are member states' preparedness, cooperation between them, and the development of security culture across the more vital sectors of economy and society (critical infrastructures).

In this sense, awareness initiatives are an essential part of the global strategy. Besides a sanctioning regime, the wide diffusion of the concept of empowerment of organizations (accountability) has forced firms to implement measures to handle cybersecurity risks appropriately.

Unfortunately, the effects of cybersecurity education are not immediate, differently from technical measures which ensure a short-term result, but as discussed in this book, are not resolutive.

Investing in people's training contributes to the construction of a cybersecurity culture, but precisely because we are talking of culture, we need time and appropriate methods and tools. The challenge is to transform human beings from weak to strong links, becoming a powerful firewall and a solid line of defence in their organization. We think that employees' education is the best key to win this challenge.

Developing awareness adequately requires understanding its meaning and its relationship with training, knowledge and education. These notions are not interchangeable: they are intertwined, but they have a different meaning and, consequently, different tools have to be applied. We can have knowledge of a certain situation, but we become aware only when we act attentively on the basis of this knowledge. We therefore denote with "awareness" something more than just knowledge, considering part of the concept both the cognitive and behavioural components. There are several examples showing how individuals do not always assume safe behaviour despite their knowledge of the risks: people drive without the seat belt fastened putting a risk their own safety or smoke intensely harming their health, even though in both cases they probably know the negative consequences.

Awareness, training, and education (defined in Table 6.1) are the three key areas of the learning process (Peltier 2005; Wilson and Hash 2003). With awareness we refer to having knowledge of a certain situation and behaving accordingly. Training is an active and a more or less formal process to teach skills. Finally, education provides knowledge and skills within a theoretical framework integrating them.

From our definitions, it is clear that the simple act of informing people, that is using materials like videos, stickers or posters to let employees know about certain issues (Spurling 1995) is just a first step in the awareness building process. It is equally clear that the only act of providing information about security risks is not

[1] The Directive on security of network and information systems (NIS Directive) https://ec.europa.eu/digital-single-market/en/network-and-information-security-nis-directive. This directive gives several obligations to European member states, for example: to adopt a national strategy, establishing security and notification requirements for operators and suppliers; designing national competent authorities and Computer Security Incidents Response Teams (CSIRTs). The directive was adopted by the European Parliament on 6 July 2016 and then transposed into national laws.

**Table 6.1** Our definitions

| Awareness | Training | Education |
|---|---|---|
| Having knowledge of a certain situation and behaving consequently. Awareness is not just having information, but developing sensitivity to important issues in an area and act accordingly | An active and a more or less formal process to teach specific skills, i.e. the "know how" knowledge allowing people to solve operational problems in a known context | The process of providing integrated knowledge and skills within a theoretical framework, which allows learners to use them to face future or unknown situations |

enough to change people's behaviour (Bada et al. 2015; Corradini and Nardelli 2019). Developing Cybersecurity Culture requires a more complex approach aiming at changing people's mindset, emphasizing the responsibility of users in information security (Ahlan et al. 2015). In short, this means growing full awareness of one's own role in security, through appropriate training and education.

To this goal, we should not forget the importance of involving employees actively, stimulating their motivation, since this affects the willingness of workers to attend cybersecurity training programmes (Noe and Wilk 1993). The question is how to motivate employees, finding the proper way to satisfy their needs (Lindner 1998; Tabassi and Bakar 2009): participation, recognition and team belonging are powerful motivators.

When employees realize that their involvement contributes to the success of training initiatives, they are recognized for their effort (not only through financial incentives) and they experience a sense of belonging, they are more likely to participate in company programmes and to give a positive input. In short, when organizational culture is more sensitive to employee's needs, it can increase their motivation (Yusof et al. 2016).

It is therefore obvious that a full consciousness in security passes through employees' engagement and motivation.

## 6.2 Tools for Awareness Development

Tools available for providing information on cybersecurity issues are the most varied, both in print and online. The current tendency is to rely on digital tools, since their production is typically less expensive and their scalability is very high, compared to printed materials.

Digital tools are also preferred because they best suit the current approach of communication strategies, which are more and more organized with digital instruments, both internally and externally. Using this approach, companies have improved some business processes and developed more collaborations between workers and management (Harvard Business Review 2016).

Regarding cybersecurity issues, the choice of communication tools depends on several factors, such as targets, type of messages, logistical aspects. Considering the different impact that the various tools can have, it is difficult to identify the best one, even though a combination of them is usually the best solution, because of the advantages and the disadvantages of each one.

The foremost action for developing awareness is to overcome the misinterpretation of cybersecurity concepts as something of interest only to IT technicians and to promote a common language, understandable by everyone. For instance, when among employees there is confusion between cybersecurity and cybercrime or among other terms, it is necessary to provide the proper meanings using communication styles adapted to the audience. In this sense, a glossary of the main terms relevant to cybersecurity and used within organization, can be a useful tool to acquire and share a common language about technical terms. Indeed, not everyone is familiar with these notions.

Discussing a glossary with employees can be an excellent exercise to explore their sensitivity to security issues and to encourage their involvement in cybersecurity awareness programmes. A glossary can be available both in print and online. Moreover, the discussion about the meaning of technical words can be used as a practical activity in specific workshop and training sessions.[2]

Some methods of delivering security information are more immediate than others and can reach a larger audience, without requiring a particular effort in designing them. Their effectiveness depends first of all on the quality of the content and then on how this content is presented in the materials used for information activity. If well done and associated to well-planned training activities, employees will receive a true benefit.

We now present the main features of different tools (summarized in Table 6.2) which can be used for developing awareness (Khan et al. 2011; Setola et al. 2015).

Among printed tools, posters can certainly reinforce some security messages through slogans able to grab people's attention even in informal areas, like cafeterias (Abawajy 2014), and they can reach company staff and external partners visiting the organization. This can have a positive effect for other guests showing how organization pays attention to security issues and, above all, to their staff preparation. The continuous exposition to posters is a good action to promote knowledge, but the permanent visibility in the same places, and the natural loss of the novelty effect, can also induce employees to ignore them.

Because of the lack of interactive contents, paper-based materials are passive tools, compared to others, which give the possibility of interaction. Drawing employee's attention is instead fundamental to avoid boredom and lack of interest. Overcoming this limit takes a good deal of creativity, an eye-catching graphical design, and an appealing communication. Considering the pervasiveness of digital tools, we might

---

[2]The author has realized some workshops with the goal of analysing employees' comprehension of cybersecurity terms used within organizations. In many cases, these terms appeared unclear and too technical or they were misunderstood.

**Table 6.2** Awareness tools

| Awareness tools | Examples | Strengths | Weaknesses |
|---|---|---|---|
| Paper-based | Poster<br>Brochure<br>Booklets | They do not require high efforts for their reading<br>Easily delivered (also to external partners) | Passive tools<br>Require update and new print (affecting the cost of their realization)<br>The amount of information to convey is limited |
| Online | E-mail<br>Newsletter<br>Messages on screen saver<br>Security tips<br>Intranet<br>Videos<br>FAQ pages | Not expensive production<br>Reach a large audience<br>High scalability | Some of them are particularly invasive (e.g. screen saver and emails) |
| Events | Annual events (e.g. security day) | Involve large audience (up to the entire organization)<br>Promote the exchange of experiences | High costs and efforts for planning |

instead think that printed tools are the real innovation; browsing the pages of a booklet or a brochure—a rarity nowadays—can be even perceived as a relaxing experience.

Online tools have the advantage of being delivered to a vast population; however, the question is not how many people receive security awareness materials, but how many of them read them, and do it actively. Indeed, employees are often oversaturated with commitments and deadlines and receive numerous emails during working hours.

Similar remarks regard the use of newsletters. This tool is helpful to spread different types of information about the company; in this sense, it can convey various security news at the same time. The problem is that people have less and less time to read; therefore, the success of the use of this instrument is subordinated to its frequency and to the amount of information to be disseminated.

Security messages can also be delivered remotely—for example through screen savers—by the system administrator in the form of slogans; but this way of diffusion can be perceived by employees as an intrusion, and consequently developing negative feeling towards these messages.

Security tips delivered to the recipients (even by email) are another means to remind employees of the proper behaviours: a list of recommendations in order to protect devices and information, like changing passwords, using VPN connections when accessing a public or unknown Wi-Fi, and so on. In spite of the importance of this type of contents, generally people are not excited in reading something describing what they should and should not do. In this case, as well as for company's policy, the best way to have an active involvement is to discuss and share the contents with the parties specifically interested, i.e. employees.

Security messages can be rapidly transmitted through the Intranet, with the advantage of being read by all personnel. The messages have to be placed so as to be visible and easily found, otherwise the risk is that individuals focus their attention on other subjects.

Security FAQs can guide people to solve problems or provide information; they need to be designed in order to cover all the questions that may interest employees and they have to be periodically updated.

Videos provide a concrete representation of the cyber-threats and the relative consequences. The wide availability of videos on the Internet, pertaining to real events, is a positive factor to promote awareness, since the message conveyed can have an immediate effect. Videos can also be included in training courses to reinforce messages or to catch attendees' attention.

There is so much material and videos available on the Internet about cybersecurity that it is possible to find videos particularly suitable for cybersecurity awareness education.[3] Videos can also be produced internally of the organization, but this is rather expensive, also considering that these products cannot be shown to the same audience repeatedly and that to be really effective they need an accompanying discussion.

Finally, in this collection of security awareness methods, we can also include specific one-day events (e.g. security day) that organizations arrange to disseminate security culture. These occurrences require significant efforts to plan them, and their effectiveness depends on a large participation of the target population. In this sense, the presence of testimonials can be a key driver to encourage people to participate actively. To be more effective, these events might be planned in conjunction with international cybersecurity awareness campaigns, in order to give more meaning to the initiative.

## 6.3 Physical Rules for the Digital World

We are used to thinking that in the digital world things work differently. This is partially true, but some rules or habits affecting our behaviour in the physical world can be applied to the digital one too.

We can take inspiration from what security experts call "cyber-hygiene", a broad concept referring to several activities to improve cybersecurity.[4] Despite this term being frequently used in both academic literature and in other contexts (media, legal, corporations), there is not yet a commonly accepted definition (Maennel et al. 2018).

---

[3] See, for example, the popular video where a clairvoyant invites young people pretending to read their minds and discovering confidential information, revealing—at the end—social networks like Facebook as sources. https://www.youtube.com/watch?v=F7pYHN9iC9I

[4] Besides cyber hygiene other similar terms are used, like digital hygiene and cybersecurity hygiene.

On one hand for some authors cyber-hygiene refers mainly to the application of practices and standards (e.g. Kerfoort 2012; Kirkpatrick 2015); on the other hand, there are those who are focused on individual behaviour (e.g. Sheppard 2013; Pfleeger et al. 2014). Finally, in describing what cyber-hygiene means, some authors prefer to consider both behavioural and technical aspects (e.g. Dobbins et al. 2015).

Regardless of the different approaches to cyber-hygiene, secure behaviour is however an integral part of Cybersecurity Culture and needs to be supported by adequate tools and techniques. In this sense, simple recommendations commonly adopted in the physical world can also be a source of inspiration for the digital space.

An example is represented by hygiene rules, well-known in everyday life. They are principles that people already know and considered good habits, transmitted through an educational process.

Hygiene rules have been taught since our childhood and they are one of the prevention measures that have most contributed over the last two centuries to lengthen the average life span. In the mid-nineteenth century, the discovery by Hungarian physician Ignaz Semmelweis was crucial for the sake of thousands of women's lives. He observed that, washing his hands after an autopsy and before assisting women in labour, the frequency of their mortality was enormously reduced. Hence, washing one's own hands became the fundamental ritual before starting surgical operations. This didn't happen overnight, though, since it took a few decades—as is sometimes the case—for this discovery to be accepted and spread.

What is more, it became one of the basic tenets taught to all children: "Wash your hands" is still the rule before going to the table, as well as "brush your teeth" before going to bed.

These rules induce people to become aware that "out there" there is an environment that can be hostile, because of microbes and infections, and that it is important to take appropriate countermeasures.

The same can be said for the digital world and therefore an analogous set of rules can be defined. This analogy is deeply true: the digital world is populated by "life forms" that are not always benign towards our "digital self" (Corradini and Nardelli 2017). Viruses and worms, for example, continue to spread at an alarming rate, and current precautions do not seem to be sufficiently effective.

The issue is that while in the physical environment this message is widely accepted and socially shared, the digital version of this message is less obvious. For example, how many people continue to insert unknown USB sticks in their computers or to open unknown attachments on their devices (PCs, tablets, smartphones) without considering the possibility that they can be infected?[5]

A partial justification is that the evolution of these digital organisms has proceeded at an elusive speed for humans. Moreover, for the traditional physical hygiene rules the current generation has benefitted from their parents' education. Similarly, we think that, even though for digital world we have not yet developed the adequate

[5]For the 2019 cybersecurity ECSA campaign, the video launched—Stay secure online—invited people to take some simple actions for their online safety, similarly to what they usually do in everyday life.

"sensors", digital hygiene can have positive effects to help people to risks and dangers in the digital environment.

Further examples based on personal protection habits can be useful to stimulate the adoption of precautions in the digital world. For instance, to travel safely, it is important to be informed about our destination. Being aware of the environment, in fact, will help us to move responsibly and consciously, reducing unpleasant risks. Of course, the unexpected can always happen, and will have to be handled (even in this case being informed helps); however, we can say that we have done our best.

Going back to the example of travelling safely and comparing with the digital world, it is more likely that familiarity with the online environment will provide us the opportunity of understanding its dangers and managing them accordingly. For a conscious use of digital technologies, it is fundamental to know how they work, and this is particularly important for young people, since they often use digital technology without knowing how they work. This is the reason why, also considering that children access online activities at an early age, it is wise to start educational actions since primary school, aiming at developing students' awareness in using digital technologies (Corradini and Nardelli 2018, 2020). In this way, in fact, they can become more confident within the technological environment and enhance their awareness of the risks we are all exposed to—with different levels of consciousness—in the current changing digital society. Moreover, this can represent a fundamental step for cybersecurity education.

Digital hygiene rules can be annoying, and even limit our range, but they are needed to safely stay and act both in the digital and physical world.

## 6.4 Preventing Failures

Despite many awareness initiatives in cybersecurity, they do not seem to be able to transform companies' staff from weakest link to "security hero" (Pfleeger et al. 2014).

The choice of the appropriate methods and tools—also discussed in Chap. 7—together with the lack of a continuous education programme on security, is one of the reasons for poor results. Furthermore, several implications should be considered when planning cybersecurity awareness programmes, in order to avoid their failure. Some of them are discussed in the following.

### *6.4.1 Security Training is Not a Cost But an Investment*

Companies should realize that cybersecurity is, above all, an issue which involves people; therefore, investing in knowledge and skills is fundamental. Updating technological measures is necessary, but if they are not supported by prepared employees, they cannot give positive results.

Enterprises should be aware that training programmes are not a cost but a priority investment, because education generates benefits for both employees and organizations. In fact, training develops employees' skills and also ensures—among others—competitive advantages, productivity, and customer satisfaction (Lashley 2018).

To understand the relevance of training, organizations should consider the costs deriving from untrained employees. For instance, workers who lack awareness can cause severe damages to their organization; moreover, when they feel unappreciated and loose motivation, they hinder the development of organizational culture and affect other colleagues negatively.

Investing in employees training means recognizing their value and their potential contribution to the success of the organization. If companies do not change their belief about the importance of training as a means to recognize people's value, this can be an obstacle for the building of Cybersecurity Culture, and the risks failing from the start.

### 6.4.2 *Training for "Troops", Not Only for "Generals"*

Despite the concerns of companies, cybersecurity training is still not a priority for them; indeed, the level of cybersecurity training is rather low.[6]

Most training activities are addressed to IT specialists and security professionals, while there is only a partial involvement of employees.[7] This represents a barrier for the development of effective cybersecurity awareness programmes: a relevant part of the organization remains extraneous to training on security problems, so people are induced to think that this subject is not part of their responsibility.

Preparing senior directors or IT specialists and not the entire population of employees is equivalent to prepare a war by training Generals, and not Troops. Then, we should not be surprised if results do not match our expectations.

For different reasons, from the role played in the company to the lack of awareness, employees are targeted by cybercriminals, who are always looking for weaknesses to be exploited. The key point is that everyone can be a target and for different reasons: for example, human resource departments are targeted by cybercriminals, because they manage significant information about employees (Verizon 2018); in the same way, new hires are a target, since they are still not aware of the functioning and of the dynamics of their organization.

[6] According to the Cyber Security Breaches Survey 2019—a quantitative and qualitative survey of UK businesses and charities conducted by the UK Department for Digital, Culture, Media & Sport in cooperation with Ipsos MORI and University of Portsmouth—only 27% of business and 29% of charities had implemented cybersecurity trainings in the previous 12 months. https://assets.publishing.service.gov.uk/government/uploads/system/uploads/attachment_data/file/875799/Cyber_Security_Breaches_Survey_2019_-_Main_Report_-_revised.pdf.

[7] According to the report cited above, directors or senior managers are the most likely staff group to receive training (81%). On the contrary employees, are among the least likely (29%).

### 6.4.3 Changing the Vision About Security Training

Companies typically tend to implement training programmes when these are mandatory or, in the absence of an obligation, when an incident occurs and an immediate intervention is needed. Such an approach is not functional to the building of a Cybersecurity Culture, because it is evident that training is perceived as something that must be done and not as an essential element to enhance organization's competencies. Therefore, before asking people to change their behaviour, organizations firstly should give a good example by changing their approach to security.

Discussing obligatory training, we can say that if on the one hand the requirement pushes companies to plan educational initiatives, on the other hand this seems to be in contrast with the principles which inspire Security Culture; indeed, this should be more based on the sharing of attitudes and values, rather than obligations.

In addition, being compliant does not necessarily mean that employees are aware of the risks and that they automatically assume safe behaviour. For example, in the area of Occupational Health and Safety (OHS), the employer does his due diligence to comply with relative regulations: employees must attend specific courses to learn to recognize risks and hazards in their workplace so as to handle them properly. However, the effectiveness of a training course depends on many factors, such as motivation, management support, job related factors, training style and environment (Punia and Kant 2013). Therefore, to achieve successful outcomes, it would be better to focus on factors affecting training effectiveness rather than on formalities.

### 6.4.4 Training as Opportunity for Employees

If many companies still consider employees' training as a cost and not as an investment, another issue is related to how employees perceive training initiatives: it is not rare that they evaluate them as a waste of time, instead of an opportunity for their individual and professional growth.

In a way, the two aspects are intertwined. When companies invest in training, they have to be aware of critical factors affecting employees' engagement, for instance the absence of a clear signal from management, poor communication, inadequate or boring methods and tools, etc.

On the other hand, when employees do not feel actively engaged, they do not attend education initiatives with the right spirit. Instead, their involvement and positive feedback help the on-going training process, since other colleagues will be stimulated to participate.

Another important issue, able to affect employees' engagement, regards the intensity of training. In an ever-changing world where there is always something we should learn, training is perceived as the main means to bridge the gap between knowledge and skills. Security is no exception and this means that cybersecurity training initiatives are deployed together with others.

All these activities require a good deal of commitment; employees are asked to carry out their usual job and additional tasks. From this point of view, over-training can be a further stress factor. We should not be surprised if the continuous demands lead employees to feel stressed and not encouraged to participate in training programmes.

### *6.4.5 Tailor-Made Training Programmes*

The success of training initiatives is subordinated to employees' participation, to their reactions and to their real behaviour change.

Two issues are fundamental in developing training.

The first is that, talking of Cybersecurity Culture, training programmes should be designed according to the values and the culture of the specific organization. A generic programme "without a soul", and unable to reflect the real values of the company, can be perceived by employees as something impersonal, and far from the work environment they experience.

The second issue is that different paths have to be designed for different recipients, keeping in mind their specific needs. In this sense, an assessment of the level of knowledge of participants (the security maturity level of the organization, Sect. 4.4) is required, so is the use of adequate contents and delivery tools. From this perspective, combining different training approaches (e.g. conventional methods and e-learning) is the best way to respond to different participant needs and to exploit the advantages of each one (Chap. 7).

Large companies can count on internal resources for the design of training activities. However, considering the importance of stimulating people's attendance and of experimenting attractive training methods, the involvement of external experts working with internal stakeholders can be extremely productive. In fact, an external eye can see organizational dynamics in a different perspective, providing useful insights for the design of educational programmes.

Finally, cybersecurity awareness programmes have to be designed keeping in mind that organizations constantly change, as do their IT infrastructures and policies. Since training contents should be consistent with the current procedures, employees should be informed and educated on security updates. Practically, periodical meetings, supported by informational activity, could be planned to discuss the policy updates, together with new risks and threats to be faced.

## References


Abawajy, J.: User preference of cyber security awareness delivery methods. Behav. Inf. Technol. **33**(3), 236–247 (2014)

Ahlan, A., Lubis, M., Lubis, A.R.: Information security awareness at the knowledge-based institution: its antecedents and measures. Procedia Comput. Sci. **72**, 361–372 (2015)

Bada, M., Sasse, M. A., Nurse, J.R.C.: Cyber security awareness campaigns: why do they fail to change behaviour? In: International Conference on Cyber Security for Sustainable Society (2015)

Bauer, S., Bernroider, E.W., Chudzikowski, K.: Prevention is better than cure! Designing information security awareness programs to overcome users' non-compliance with information security policies in banks. Comput. Secur. **68**, 145–159 (2017)

Corradini, I., Nardelli E.: Digital hygiene: basic rules of prevention, in Broadband4Europe (2017). http://www.broadband4europe.com/digital-hygiene-basic-rules-prevention/

Corradini, I., Nardelli, E.: Building organizational risk culture in cyber security: the role of human factors. In: Ahram, T., Nicholson, D. (eds) Advances in Human Factors in Cybersecurity, AHFE 2018. Advances in Intelligent Systems and Computing, vol. 782. Springer, Cham (2019)

Corradini, I., Nardelli, E.: Awareness in the online use of digital technologies of Italian students. In: Proceedings of ICERI 2018, 11th Annual International Conference of Education, Research and Innovation (2018)

Corradini, I., Nardelli, E.: Developing digital awareness at school: a fundamental step for cybersecurity education. Submitted to: 6th International Conference on Human Factors in Cybersecurity, AHFE, 16–20 July 2020, San Diego, California, USA

Cyber Security Breaches Survey, UK Department for Digital, Culture, Media and Sport, Ipsos MORI, University of Portsmouth: Cyber Security Breaches Survey (2019)

Dobbins, J., Solomon, R.H., Chase, M.S., Henry, R., Larrabee, F.S., Lempert, R.J., Liepman, A.M., Martini, J., Ochmanek, D., Shatz, H.J.: Choices for America in a Turbulent World: Strategic Rethink. Rand Corporation (2015)

ENISA: Cyber Security Culture in organisations (February 2018). https://www.enisa.europa.eu/publications/cyber-security-culture-in-organisations

Harvard Business Review: Why Today's Digital Teams need a new class of collaboration tools and solutions (2016)

Herold, R.: Managing an Information Security and Privacy Awareness and Training Program. CRC Press, Taylor&Francis. Second edition (2010)

Lashley, C.: Investing in people: training Is Not a Cost! In: Oskam, J.A., Dekker, D.M., Wiegerink, K. (eds.) Innovation in Hospitality Education. Innovation and Change in Professional Education, vol 14. Springer, Cham (2018)

Leboeuf, M.: The Greatest Management Principle in the World, Putnam Pub Group (1985)

Lindner, J.R.: Understanding employee motivation. J. Ext. **36**(3), 1–8 (1998)

Khan, B., Alghathbar, S.K., Nabi, S.I., Khan, M.K.: Effectiveness of information security awareness methods based on psychological theories. Afr. J. Bus. Manage. **5**(26), 10862–10868 (2011)

Kirkpatrick, K.: Cyber policies on the rise. Commun. ACM **58**(10), 21–23 (2015)

Maennel, K., Mäses, S., Maennel, O.M.: Cyber Hygiene: The Big Picture. Secure IT Systems: 23rd Nordic Conference, NordSec 2018, Oslo, Norway, 28–30 November 2018. In: Gruschka, Nils (ed.) Proceedings, Cham: Springer, 291−305 (2018)

Noe, R.A., Wilk, S.L.: Investigation of the factors that influence employees' participation in development activities. J. Appl. Psychol. **78**, 291–302 (1993)

Peltier, T.R.: Implementing an information security awareness program. Inform. Syst. Secur. **14**(2), 37–49 (2005)

Pfleeger, S.L., Sasse, M.A., Furnham, A.: From weakest link to security hero: Transforming staff security behavior. J. Homel. Secur. Emerg. Manage. **11**(4), 489–510 (2014)

Punia, B.K., Kant, S.: A review of factors affecting training effectiveness vis- à-vis managerial implication and future research directions. Int. J. Adv. Res. Manag. Soc. Sci. **2**(1), 151–164 (2013)

Setola, R., Mastrapasqua, M., Tesei, M., De Maggio, M.C., Corradini, I., Pantaleo, C., Capitello, M.E., De Simio, F.: Study on security awareness in gas infrastructure. Final report, NITEL (2015)

Sheppard, B., Crannell, M., Moulton, J.: Cyber first aid: proactive risk management and decision-making. Environ. Syst. Decis. **33**(4), 530–535 (2013)

Spurling, P.: Promoting security awareness and commitment. Inf. Manag, Comput. Secur. **3**(2), 20–26 (1995)

Tabassi, A.A., Bakar, A.H.A.: Training, motivation, and performance: the case of human resource management in construction projects in Mashad, Iran. Int. J. Project Manag. **27**, 471–480 (2009)

Verizon, 2018 Data Breach Investigation Report, 11th edition. https://enterprise.verizon.com/resources/reports/DBIR_2018_Report.pdf

Yusof, H.S.M., Said, N.S.M., Ali, S.R.O.: A study of organizational culture and employee motivation in private sector company. J. Appl. Environ, Biol. Sci. **6**, 50–54 (2016)

Wilson, M., Hash, J.: Building an Information Technology Security Awareness and Training Program, National Institute of Standards and Technology (NIST), Special Publication 800–50, Computer Security (October 2003)

# Chapter 7
# Training Methods

**Abstract** There are many available methods to develop training on cybersecurity issues, and it is not easy for a company to identity the most suitable one. However, the choice of the most appropriate training method requires a preliminary assessment of the training needs of the company. Methods can be classified according to different criteria, such as the place where they are carried out (internally or externally to the organization), tools and techniques used, the role of the instructor, the level of interaction with participants, and so on. Over the last few years, training based on the use of technologies has been widely developed, as have methods to engage employees actively—interactive and practical approaches—like gamification and serious games, simulation, role playing. Considering the limitations of the different training methods, a blended approach can provide a balanced solution. To achieve the goal of changing employees' behaviour and their attitude towards security, training should be part of a continuous education process, and for what concerns its implementation within organizations, it should be constantly monitored. Moreover, considering the availability of innovative trainings methods exploiting digital environment, a combination between these and conventional approaches could be an excellent strategy towards security-conscious behaviour. Regardless of the approach used, the key element for its positive outcome is the ability of designing training programmes carefully and making them emotionally attractive. In this chapter we discuss different methods which can be used for cybersecurity training—identifying their points of strength and weakness—and implications for measuring their effectiveness.

## 7.1 Training Methods Are Not All the Same

The choice of training methods to educate employees on cybersecurity issues is an important step for the effectiveness of cybersecurity awareness programmes; goals, stages, tools and techniques have to be planned and coordinated in order to produce the learning outcomes that the organization wants to achieve. To select the most appropriate training method we should start from the assessment of the training needs of the company and then identify training goals and targets.

© Springer Nature Switzerland AG 2020 
I. Corradini, *Building a Cybersecurity Culture in Organizations*,
Studies in Systems, Decision and Control 284,
https://doi.org/10.1007/978-3-030-43999-6_7

Both conventional and experiential approaches, as well as online courses, have strengths and weaknesses; however, regardless of the chosen method, what makes the difference is the accurate design.

Over the last few years, training based on the use of technologies has been widely developed. Considering the familiarity that people have with digital technology during their daily activities, the use of online training methods could ensure a better engagement of employees in the training programmes. In fact, given the various opportunities offered by digital technology, people could also be willing to be educated by way of their smartphone. Online training answers the needs of the digital era and, at the same time, provides some benefits; for instance, it minimizes the costs deriving from in-person training, like travel and accommodation, or of the time spent away from work. Anyway, this kind of training is not always so effective.

We are fully aware that companies are particularly sensitive to costs, and training is not an exception. On the other hand, they need to invest in updating employees' skills, and for a large company with thousands of employees this means a significant investment. It is therefore easy to understand the reasons of the wide use of e-learning courses: they are easily scalable and their application is not complicated, since it just needs a computer (or a digital device) and the right content. Precisely because it is online, this approach requires an extremely accurate design in order to be appealing for the audience and effective in terms of outcome. This is especially true considering the security area, which by nature is often perceived as boring or something belonging to the field of security experts.

In any case, companies should focus their efforts on achieving a successful outcome in protecting business, going beyond a solely economic perspective. Sometimes, companies spend lots of money without achieving what they expected. Time available for implementing such initiatives can be an obstacle to their effectiveness; unfortunately, for many companies, time devoted to train employees—especially on security issues—is not always seen favourably. They are convinced that a training course is enough to face security problems, when it is evident that the change of behaviour is not immediate and requires continuous education actions.

The size of the organization and, above all, its level of cybersecurity maturity, drive the choice of methods and techniques to be used. For instance, a large company will probably prefer online methods, in particular when their employees work in other countries, or when other training actions have already been developed. A small company, on the other hand, has the possibility of adopting a face-to-face approach.

Training methods can be classified according to different criteria, such as the place where they are carried out (internally or externally to the organization), tools and techniques used, the role of the instructor, the level of interaction with participants, and so on.

Since many methods are available, it is not easy for a company to identity the most suitable one. In the following we describe some methods that can be applied for cybersecurity training, distinguishing them between conventional, experiential and informal.

Keeping in mind the goal and the nature of the subject, interactive and practical approaches are strongly recommended. The preference for these methods is essentially related to the nature of the audience; indeed, according to the principles on adult learning education (Knowles 1980, 1983)[1] the role of experience is fundamental, because it allows to anchor learning to problems and situations perceived as significant: individuals are available to learn what they need to face life situations.

Besides methods, it is fundamental to develop adequate contents answering the goal of the training. If we want to develop a different mindset on security, the contents cannot be presented as a "to-do" or "not-to-do" list; instead, they have to stimulate curiosity and induce individuals to ask themselves what would happen if certain situations involved them.

In this sense, topics to be included in the training should not just consider technical questions, but they should connect them to the human experience: for example, explaining what social engineering is cannot be effective if it is not supported by real life examples discussing how psychological mechanisms work and what human beings can do to prevent this kind of threat. The principles of persuasion discussed in Chap. 3 are a helpful resource for setting examples pertaining to the specific organization.

Finally, including real life cases, also relating to individuals' private life, provides the opportunity of drawing employees' attention; they could be stimulated to reason on other benefits, for example extending the knowledge of good practices in security to members of their own family (e.g. spouse, children).

## 7.2 Conventional Methods

Among these traditional approaches, instructor-led classroom is one of the most popular. It is focused on the role of the trainer, on his public speaking skill, and on his ability to guide learners. Conventional methods are typically developed face-to-face, differently from online-based training.

On the basis of the level of interaction planned by the teacher, they can reach both large and small audiences physically present in the same place. Teacher-classrooms, just like seminars, can be expensive to plan and run; they require a logistical organization, especially when companies need to replicate training sessions for a large number of employees operating in multiple locations.

The trainer can use slides, photographs, and videos to encourage questions from the participants; his role is significative because he has to handle all classroom dynamics, and the success of the training depends on his competences and public speaking. This is why the choice of the instructor is a priority for the effectiveness of this type

[1] Knowles popularized the concept of "andragogy", involving learning adults, in contrast with "pedagogy", focused on children education. According to his approach the individual's experience and active involvement in the formative experience are fundamental. Six assumptions are fundamental in this approach: need for knowledge; self-concept; role of experience; willingness to learn; learning orientation; motivation.

of training; subject knowledge is not enough, it is also important to have the ability of communicating the contents. This is an element which should not be neglected, since—unless the audience is made up of technical and specialized individuals—people feel far from security issues; consequently, one of the major tasks for the instructor is to open up people to the subject. In addition, if on one hand some questions require specific knowledge on cybersecurity (e.g. the details of a cyber-attack), on the other hand, we have to consider that if we want to act on individuals' mindset, we need an approach more empathic and closer to learners' lifestyle.

Even though traditional methods have a limited interactivity, the trainer can expand it: in fact, the physical presence of learners in the training location represents a practical opportunity for fostering helpful discussions with the expert and among participants, as well as sharing their experiences.

Another way to integrate conventional methods with interactive techniques is to rely on working groups, for example assigning them the analysis of a case study on cybersecurity. In this situation each group first discusses internally to the group the proposed item, then together with the trainer and the other groups.[2] It is obvious that the introduction of working groups in the traditional lesson is subordinated to the number of participants, considering the difficulty for a single trainer to handle many groups simultaneously.

Combining the traditional lesson (theoretical part) with an experiential practice offers several benefits to participants, since it is possible at the same time to transmit contents, to share experiences and overcome the risk of boredom.

Finally, testimonials working in cybersecurity field might be invited during the classroom training to reinforce some positive security messages and discuss successful case studies. This kind of experience can be an excellent incentive for participants.

## 7.3 Experiential Learning Methods

These methods ensure a greater involvement of participants, because they emphasize experiential activities (hands-on) supported with reflective observations (Young 2002). They are able to overcome the limitations of one-directional knowledge characterizing the traditional teacher-classroom method, and anyhow achieve the goal.

The basic assumption is that hands-on experience is able to provide a more effective learning, since it helps learning by restructuring and systematizing concepts and attitudes. In particular, learners engaged in problem solving activities increase their motivation and are induced to think and use information in more productive way (Greitzer et al. 2007).

[2]This approach has been used by the author in many training courses on cybersecurity issues. For the organization of the working groups it is important to identify the criteria for their constitution; for instance, the heterogeneity of the group is a key factor for the success of the initiative.

Even if these methods generally have a high level of interaction, we cannot take for granted the complete availability of participants, since not everyone is so extrovert to be actively engaged. It is therefore important to design the training in order to equally involve all attendees. Moreover, people have to be well aware that—even if the training programme may be presented using a ludic scenario—they are participating in an educational experience.

Several methods included in this area present some points of contact and the boundaries are rather blurred, so it is not easy to differentiate them.

Choosing among them depends on different factors, like the organizational culture, temporal and logistic aspects, etc.; however, regardless of the type of experiential method one decides to adopt, it is fundamental to ensure from the beginning a clear design, with the support of professional competences.

In any case, it would be useful to develop training activities with these methods counting on a multidisciplinary team, since the discussion of incident handling requires both technical and social skills, and different analysis perspectives.

In the following we describe some methods used-or which could be used- for cybersecurity training.

### *7.3.1 Case Study*

This method consists of exposing a case history or a real situation with the aim of encouraging the discussion among participants, promoting decision-making skills through its analysis. Starting from questions, like: "how it happened?", "what should we do or not do?" attendees are asked to investigate the facts and, on the basis of the available data, formulate proposals for the solution.

Case study represents a good exercise for cybersecurity training, since studying a real-world case of data breach (Cai 2018) participants can discuss the different perspectives and find out the best arrangement for handling it.

Considering the amount of significant data breaches happened in recent years, it is not difficult to select materials to be studied with participants. The question is whether to choose a case study applicable to the specific company: on one hand a pertinent situation makes it more real and appealing for participants; on the other hand, a case study coming from another sector can be challenging, and provides a different point of view. Moreover, this has the advantage of not putting under the spotlight facts and people internal to the organization related to the critical events. The choice of the approach is also linked to the level of security maturity of the organization and to the type of initiatives already carried out.

Using real examples, employees will have the opportunity of exploring causes and effects of a cyber incident or a cyber-attack, and of identifying the critical and positive elements of the situation presented. By doing so and working in groups, they can understand the importance of human actions and mechanisms both in the prevention and in the management of cybersecurity issues.

### 7.3.2 Role-Playing

This is one of the most known active methods of learning, deriving from the Moreno's psychodrama[3] based on the structure of the drama.

Role-playing has the goal of reproducing problems and events similar to real life where participants are asked to play a specific role. The logic is that one tries "to put oneself in the shoes of" starting from predefined role profiles.

At the end of the exercise, which develops for a limited amount of time, the moderator activates and guides the discussion with participants, in order to analyse the different experiences and the dynamics emerged during the interactions. This is why for an effective achievement, it is important to count both on the high experience of the conductor, and on the presence of one or more observers whose task is to collect as many details as possible during the representation.

Role-playing, more or less structured, facilitates the analysis of problems from various perspectives; in this sense, the inversion of roles can be used, as well their rotations, which means that all participants are requested to play the identical role. It is important to design the situation accurately and deliver in advance to the participants the necessary details to play well the respective roles; the "actors" must communicate and act following the information provided, while relying on their personal creativity to fill-in any required missing detail. If well-planned and conducted this method can be an extremely useful learning experience; however, like all training methods, role-playing requires a qualified expert able to guide the representation and to analyse the outcome.

This technique is typically applied in exploring complicated social situations and it is popular in education, but even in cybersecurity it can be very helpful for studying user's security behaviour (Yao et al. 2014). For instance, it is possible to ask participants to manage a cybersecurity incident, assigning them different roles: it will be possible to verify if the acted behaviour needs to be improved.

### 7.3.3 Gamification and Serious Games

A number of games applied in cybersecurity training has been developed over the last few years, so that gamification methods have become more and more popular, also in the eLearning experience. They take inspiration from the dynamics of problem solving and their appreciation derives from several factors: games are considered a good mean for learning (especially for children); the digital version promotes the production of reliable scenarios; finally, gamification is user-friendly.

---

[3]Psychodrama is a therapeutic model, widely used in health settings, developed by the J.L. Moreno. Based on experiential methods, like role playing, it is used to help patients work on their problems actively.

The design of a game for cybersecurity training is not a simple endeavour (Nagarajan et al. 2012) and its implementation and maintenance are not easy; however, considering the rapid evolution of the security landscape, this educational approach is likely to grow.

The participants are involved in a competition, typically divided into teams (e.g. red and blue teams), and they are requested to develop their creativity, strategic analysis and decision making. In this sense, gamification responds to the educational paradigm of "learning by doing", but actually the effective outcome of this kind of exercise is still not clear (Maennel et al. 2017). The attendees make decisions on the basis of data distributed at the beginning of the game, and then keep taking decisions derived from the previous ones. Some incentives are provided (point and reward systems) to engage the participants.

We can say that they live a psychological experience, since in the conceptualization of this approach, three fundamental parts are involved (Hamari et al. 2014): motivational affordances, psychological outcomes, and behavioural outcomes. Gamification approach encourages users to achieve a shared goal. Furthermore, the "game dynamics" induces individuals to develop relationships with others and to improve their personal abilities, especially in a challenge perspective.

The use of games for cybersecurity training requires that their design have to be planned in agreement with the specific target involved.[4] We can find several applications of gamification in the context of Cyber Defence (Amorim et al. 2013) where it is possible, for instance, to build sophisticated crisis management scenarios.

Other applications in cybersecurity training include serious games, reproducing "a mental context" (Zyda 2005) using traditional game techniques, like video games. Even if serious games are often grouped with gamification as eLearning experience, they differ because their conception goes beyond the pure entertainment and include a pedagogical element (Le Compte et al. 2015).

Considering their goal, they are typically highly-structured. For example, in designing them, there is the question of maximizing users' memorability by means of graphical and verbal cues (Micallef and Arachchilage 2017), since, according to research, human being remembers images better than words (Paivio et al. 1968).

Serious games have been implemented for cybersecurity awareness and training especially in formal contexts[5]; for an effective outcome their design should adopt a socio-technical approach (Kianpour et al. 2019), in order to include the nature of the cyber domain, the motivation of the attendees, and experiential learning.

---

[4]For example, the Game of Threats developed by PricewaterhouseCoopers (PwC) is designed for senior executives and boards of directors.

[5]See, for example (Le Compte et al. 2015) the game "CyberCIEGE", developed by the US Naval Postgraduate School, and other games, like "CyberProtect", "Anti-phishing Phil" (Carnegie Melon University).

Despite gamification and serious games are considered a frontier to be developed in cybersecurity training and exploited in international competitions, research methodology in this area requires further development,[6] both on the evaluation side and on the actual goals achieved. In fact, if games aim to change people's long-term security behaviour, it is not clear how it is possible to achieve this goal through a single intervention developed in a short period (Hendrix et al. 2016).

### 7.3.4 *Simulation*

This approach consists of the representation of some real-world system (Crookall and Saunders 1989), producing a genuine scenario both in content and in relational dynamics. Simulation offers an educational experience in a safe environment.

Despite containing rules and strategies like games (Garris et al. 2002), simulation differs from them because it does not need to represent a conflict or a competition; indeed, the identification of winners and losers does not characterize this approach (Sauvé et al. 2007).

The application of simulation in cybersecurity education can cover the most varied situations, such as responses to terroristic and cyber-attacks. In the military field, for example, this method is often used to test the resilience of critical infrastructures.

This approach can also be used within organizations to simulate a cybersecurity incident requiring executive-level decision-making and adequate communication strategies (EY 2017). These exercises can be addressed to a high target audience, such as C-suite managers and board members.

By using simulation, it is possible to evaluate the adequacy of coordination and the effectiveness of procedures planned for an emergency situation, as well as to verify the appropriateness of available resources.

In conclusion, simulation is a hands-on method where people can really experience what happens and what needs to be done to handle a critical situation. In ever more immersive way, it is finding a large application in cybersecurity training.[7] However, the construction of the scenario and its implementation can be very expensive for many organizations.

---

[6] See, for example, the Proceedings of the 5th Interdisciplinary Cyber Research Conference 2019, Tallin University of Technology (Osula and Maennel, Editors). https://www.taltech.ee/public/k/kyberkriminalistika_ja_kyberjulgeoleku_keskus/proceeding.pdf.

[7] See, for example, the initiative of Homeland Security Department (USA), which is funding a new immersive cyber-training platform equipped with simulation-based scenarios and exercises aimed at protecting the nation's energy sector. https://www.nextgov.com/cybersecurity/2019/03/dhs-invests-59-million-cyber-training-tool-energy-sector/155808/.

## 7.4 Online Methods

Over the last few years the number of online courses has increased, and e-learning market is bound to grow.[8]

More and more companies are using these methods to train their staff. In fact, especially for large companies, the online approach represents an opportunity to reach a wide audience—in particular when employees operate in multiple locations—without incurring prohibitive costs. Because of the ubiquity of digital technologies, even universities are increasingly geared to implement eLearning courses.

Accessing to online platforms (web-based training), participants can define personal learning paths (self-paced learning), also with the help of a facilitator (instructor-led training). Regarding enterprises, online courses are hosted on their intranet or on specific platforms often constructed in cooperation with outside specialized firms.

E-learning has several advantages: it can disseminate the same contents to a large audience and it can be attended at any time and by using various devices, so offering a flexibility suitable to the different commitments of employees. Resources are available in varied formats and can be seen/heard repeatedly.

There are many online cybersecurity courses, free and professional, also addressed to companies to develop their programmes for IT specialists and employees. On the Internet it is possible to find several websites offering training on the most various themes, from cryptography to social engineering attacks prevention. However, to meet the specific requirements of the company, these courses should be tailor-made and adapted to the actual context.

Even though online approach is considered economically convenient compared to traditional methods, there is no consensus on its effectiveness. In addition, it is difficult to discuss this concept for e-Learning (Noesgaard and Ørngreen 2015). To give positive results, online training should count on several factors, such as a good deal of interactivity, a solid plan of communication and, above all, a highly motivated audience. Its contents need to be presented in an engaging way: the challenge is to avoid being perceived by employees as a boring experience, thus failing to achieve the expected outcome.

Besides the contents, the duration of training is important, because of the risk of losing learners' attention. For this reason, interactive modules should be included; moreover, the use of graphics should be bound to deliver information more effectively and not to be an element of distraction.

Good designed e-learning courses require time and effort for their preparation, but they allow employees to decide the best time to participate in the training initiatives. However, considering that companies always have new needs and little time to update their staff, they are increasingly adopting digital learning in the broadest sense of the word, going far beyond the traditional e-learning. In this sense, a set of channels and

---

[8]According to e-learning Market–Global Outlook and Forecast 2019–2014 Report, the global e-learning market is expected to reach $ 238 billion by 2024 https://www.reportlinker.com/p05289135/E-learning-Market-Global-Outlook-and-Forecast.html?utm_source=PRN.

tools are used with the goal of supporting and strengthening training programmes: webinars, social networks, mobile applications and several informal methods (see the Sect. 7.5).

By so doing, the educational approach can be used synchronously or asynchronously and implemented in different ways according to the needs of users, overcoming the lack of interaction of the trainer. For example, webinars provide a nearly face-to-face environment able to promote a multi-level interaction (Wang and Hsu 2008).

## 7.5 Formal Versus Informal Methods

Even if formal learning (discussed in the previous paragraphs) is considered one of the most reliable methods to enhance people's knowledge and skills, other forms of training can help this goal, responding to specific needs, such as time restriction and developing positive relationship at work.

Educating people is an on-going process, a continuous nurturing; therefore, further informal training initiatives are encouraged, even though these are not a replacement for formal training. At the same time, living in a digital society requires adapting to new learning methods and finding different ways to stimulate individual's motivation to learn.

We can therefore consider other forms of training based on learning deriving from human interaction (e.g. coaching and mentoring), or on a spontaneous learning directly driven by users (informational learning). In the following we briefly describe the two approaches.

### *7.5.1 Coaching and Mentoring*

Coaching and mentoring are often used interchangeably; actually, these methods have both similarities and differences.

Coaching can be addressed to individuals—and also groups—with the aim of developing their potential in order to achieve specific goals through an autonomous process of learning (Whitmore 2009), where the coach has a role of facilitator. Despite this method being widely known, there is still not an agreement on its definition. In fact, coaching is often confused with analogous terms and overlapped with other disciplines.

On the basis of the current literature, coaching can be discussed from several perspectives (Ives 2008), for example humanist, behaviourist, cognitive, goal-focused, system-oriented, each one with their specific goals. Differently from the traditional training, the program is chosen by the coachee; in addition, coaching is performed on site and highly personalised.

The word "coaching" evokes the image of a sport context. Looking at its application, indeed, we can find successful experiences in supporting tennis players (Gallwey 1975); this approach was then extended to work and training.

Since coaching aims at changing dysfunctional habits and implementing positive behaviour, and given that human behaviour is a theme of interest for cybersecurity, it may be worth evaluating the application of this approach for behavioural change in cybersecurity. Coaching is typically used for managers, but it can be addressed to whoever is interested in improving his personal performance in various areas and increasing individual motivation.

We have already stressed how important the involvement of top management is for the success of Cybersecurity Culture; keeping in mind its pressing work commitments, coaching might respond to their specific needs.

Moreover, as discussed in this book, executives are themselves a priority target for cybercriminals; consequently, they need to be trained—just like their employees—and for this goal individual coaching sessions on cybersecurity could be planned.

When seniors take the informal role of guiding juniors we can talk of mentoring. This approach is based on a personal relationship developed—formally or informally—between two actors: on one hand a person with more experience (mentor), on the other hand an individual with less experience (mentoree). Mentor is perceived as a role model, a counsellor (Anderson and Shannon 1988), who supports the inexperienced person in developing skills useful for task performance and social life. Differently from the informal version, formal mentoring takes place within a defined process and program; in terms of effects, it seems to be less effective than informal mentoring (Ragins and Cotton 1999).

In the work environment coaching and mentoring can develop similar activities, to the point where a mentor can assume both roles (Abiddin 2006). Actually, several differences separate mentoring from coaching (Minter and Thomas 2000). The main difference is related to the teacher expertise: the mentor is typically experienced in a particular field and has the technical knowledge necessary to perform the specific job of the mentor. The coach, instead, has the role to help the coachee to overcome problems and lead him to find the solution by himself, regardless of his specific expertise.

### 7.5.2 *Informational Learning*

Other forms of learning activities can be independently chosen by individuals to increase their knowledge and skills. These activities can be conducted alone or include the participation in informal groups, like friends and colleagues. Searching a topic online, reading a white paper, attending a webinar or a conference are examples of actions which can be experienced to deepen the knowledge of particular issues. For webinars and conferences, there is also the opportunity of discussing the contents of training with others. Experts often use this type of activities for their professional development.

However, informational learning can involve everyone, since digital tools provide learning opportunities for all users. In fact, in a context characterized by an increase of available information, it is normal for people to try new ways of learning. Searching on the Internet represents the easiest way to be constantly informed, even though this does not mean to be correctly informed. It is also possible to participate in forums and projects, watching videos, reading articles and papers on the Internet through own digital devices. The main advantage is that these activities can be done anytime and anywhere, and digital communities become a learning setting, overcoming the physical locations in which learning occurs.

In this sense, informational learning matches with the philosophy of lifelong learning, a continuous individual's learning process promoting the development of knowledge and practices. This approach goes beyond school and professional life, and it is directly driven by learners, who can decide the best course of action. From this point of view, learning can be perceived as a positive and relaxing experience, since it is not subordinated to exams, dates, and all those factors that typically make training stressful.

Even if this type of training can be seen as a spontaneous method, it is clear that its success is strictly related to individuals' engagement and motivation.

## 7.6 Blended Approaches

Considering the limitations of the different training methods, a blended approach can provide a balanced solution. It this way it is possible to combine, for example, e-learning sessions with face-to-face training initiatives. This "hybrid" approach can answer the different needs of organizations, especially the larger ones, and be extremely flexible. The combination of various approaches can also be perceived positively by employees, especially by those who do not really appreciate traditional training.

Their application requires an open mindset, able to overcome prejudices related to the adoption of a specific way of training. Moreover, it is necessary to carefully evaluate when and how to apply the different methods (e-learning or face-to-face), keeping in mind the contents to disseminate and the targets to involve. Finally, to adopt such an approach successfully, organizations have to ensure a proper coordination of all activities implemented.

In the following table we summarize the training methods discussed in this chapter, identifying strengths and weaknesses (Table 7.1).

## 7.7 Measuring the Effectiveness of Cybersecurity Training Programmes

Evaluating the effectiveness of the training activities for building Cybersecurity Culture is a fundamental step to evaluate the appropriateness of the initiatives carried out and, where necessary, to provide corrective actions (Corradini and Nardelli 2019;

**Table 7.1** Types of training methods

| Methods | Examples | Strengths | Weaknesses |
|---|---|---|---|
| Conventional | Instructor-led classrooms, seminars | In person training<br>Addressed to both a large and a small audience<br>Direct human interaction encourages discussion | Expensive (location, outsource trainer, replication of training sessions, etc.)<br>Limited interaction<br>Success depends on the ability of the trainer |
| Interactive and experiential | Case study, role-playing, gamification, simulation | Participants are actively involved<br>They live a psychological experience<br>Activities stimulate interests and promote sharing ideas | Time-consuming for their preparation<br>Individual differences to be considered in the training design<br>Evaluation of the effectiveness is not clear |
| Online | Web-based training, self-paced training, instructor-led training, webinars | Reach a wide audience, also in multiple locations<br>Address the same contents to a vast population<br>Resources are available in various formats<br>Participants can study at their own pace and location | Time-consuming for their preparation and expensive<br>Design should include interactive modules and appealing graphics<br>The effectiveness of this approach is not clear |
| Informal methods | Coaching, mentoring, | Learning derives from human interaction<br>Deployment of these methods can be formal or informal | High-costs for coaches (especially if external)<br>Lost working hours<br>Time to develop high quality mentors |
| | Informational learning | Driven directly by learners<br>Variety of activities | These methods are not a replacement for formal learning (they support them) |
| Blended approaches | Combination of different training methods | High flexibility<br>Answer different needs of the organization<br>Overcome the limitation of single methods | Their application requires:<br>– an open mindset<br>– a careful evaluation<br>– a proper coordination |

Rantos et al. 2012). However, the process of measuring cybersecurity awareness programmes is not an easy task, given that we are talking of the real change of people's behaviour.

The evaluation of the outcome is linked to the design of the programme, since it is possible to measure its impact on people's behaviour when we already have a preliminary assessment. For example, starting from the assessment of the level of security maturity of the organization, it will be easier to identity the results achieved after the actions taken.

Here, we mainly focus on the effectiveness of cybersecurity training, but it is evident that a global evaluation of a cybersecurity strategy should also include other elements, such as technology security standards and processes.

Before discussing methods and outcomes, it can be useful to distinguish between the concept of measure and metrics (Black et al. 2009): a measure is an objective evaluation of a given physical quantity, while a metric is the subjective choice of a set of quantities that we think may provide information on a situation. By collecting measures, we obtain an evaluation of the current level of the metric.

Metrics can have the goal of verifying the real change of employees' behaviour or to understand the impact of the cybersecurity programme on company business. In the first case metrics can include, for example, the number of workers victims of phishing emails, or consider all those who use weak passwords. In the second case, instead, it could be more helpful to calculate the number of cyber incidents occurred or the average time taken to detect them. However, the issue of identifying security metrics and measuring security programmes is very complex (e.g. Payne 2006; Chew et al. 2008).

In choosing the approach to measure the effectiveness of cybersecurity awareness programmes, we can proceed in different ways (Enisa 2018):

- take a picture of the current situation evaluating employee's awareness and remeasure it after the implementation of training initiatives;
- create a list of metrics (e.g. not clicking on links appearing in untrusted emails) and implement adequate measurement methods (referring to the example cited: sending phishing emails fabricated ad hoc for this purpose);
- integrate both approaches.

Regardless of the approach selected, at least two issues—which are connected to one another—have to be discussed: what and how to measure.

Concerning "what" has to be evaluated, we have above all to separate what people have learned attending cybersecurity training and what they actually transfer into their work environment. In fact, behaviour does not change overnight or after a training session; people can change their behaviour if they are supported by a continuous cycle of training and information activities, whose effectiveness also have to be measured separately.

It is evident that one thing is to evaluate the outcome of each training session, another is to identify any consequent permanent change of behaviour.

**Table 7.2** Levels of training evaluation

| Level | Description | Methods (examples) |
|---|---|---|
| Reaction | The level of satisfaction of the course, strengths and weaknesses of the programme | Reaction questionnaires, multiple-choice questions |
| Learning | Learning of knowledge and skills | Performance tests |
| Behaviour | Actualisation of learning at work, real change behaviour | Data, observation, monitoring |
| Results | Business results, higher morale, increased production, improved quality and efficiency | Combination of methods for measuring costs, productivity, quality, ROI (return on investment) |

For the evaluation of training effectiveness, several approaches are available in literature. Among these the four-step model (Kirkpatrick 1976, 1996, 2007) is one of the most influential and commonly used. It is based on four levels:

1. opinions about the learning experience of who attended the course (reaction)
2. learning of knowledge (learning)
3. real change of behaviour on the job (behaviour)
4. impact on company's business (results).

This model has been subject to several criticisms, for example with respect to the hierarchy of values related to the different steps (Tamkin et al. 2002); other approaches, emphasizing distinct elements of the evaluation and using different measures, can be found in literature.[9]

However, Kirkpratick' model can be considered a pragmatic approach, easily understandable, and it provides helpful insights for the evaluation (see Table 7.2). For example, reaction to learning experiences can be easily investigated at the end of the course using simple questionnaires or interviews (level 1).

The evaluation of the learning outcomes linked to the training method applied is another question (level 2). Typically, for traditional method (like instructor-led classrooms) multiple-choice questionnaires are applicable, while for other types of interventions things are not so linear. For example, for games used in cybersecurity education, not everyone seems to have a measurable (or a conclusive) effect from a learning point of view, even though these methods are appreciated by participants (Hendrix et al. 2016). Therefore, more robust evaluation methods in this area are needed.

If an immediate evaluation about the knowledge achieved can be gained by the use of questionnaires or interviews, measuring the change of behaviours requires a different approach and a combination of methods (level 3). Indeed, considering the complexity of the behavioural concept, it is not an easy task to measure the real change.

[9]See, for instance, the Hamblin Model (1974), the Organizational Elements Model, OEM (Kaufman et al. 1995); the learning outcomes approach (Kraiger et al. 1993).

From a scientific point of view, the evidence of the effectiveness of changes deriving from training actions should be based on intervening factors and not on extraneous variables (Blythe 2013): in this sense, for instance, in the health and psychological domain, the scientific procedure of randomized control trial (RCT) is widely applied. However, we are aware that such an application in enterprises can be extremely complicated, because of the difficulty of defining the proper sample and the appropriate method to proceed as it would be done with a research project. For instance, not all employees participate in the same initiatives, or new employees coming from other experiences can have a different background and approach to these themes.

Considering the objective difficulties for companies and their specific needs, they can use a combination of approaches to measure the behaviour change, responding to the second issue cited above, "how" to measure.

Looking at a numerical and non-invasive approach, the reduced number of phishing emails openings or the increase in reporting them provide some information about the achievement of protective behaviour, compared to the number of openings and reports preceding the training activity. In the same way, the results of penetration tests based on social engineering techniques can provide some evidence of a change approach to security.

Despite a numerical approach can provide a reassuring quantitative measurement, we cannot be sure that people have really changed their behaviour. Considering strengths and limitations of quantitative and qualitative research (e.g. Carr 1994; Black 1999; Punch 1998; Sale et al. 2002), a combination of both approaches provides a more accurate evaluation. Hence, beyond collecting data and analysing them by means of standard statistics, the evaluation of training effectiveness should also be supported by other qualitative methods, e.g. focus groups, interviews, participants observation, open-ended questionnaires.

Monitoring and measuring people's behaviour before and after training is not to be intended as a "control activity" on people's actions, but an informal and cooperative interaction with them, in order to understand what more can be done to fill gaps. Interviews and focus groups are helpful methods to motivate people to improve or maintain their protective behaviour.

Measuring the effectiveness of a training program is essential but not conclusive; indeed, to evaluate if an adequate level of security maturity has been achieved it is also necessary to measure what people think and perceive about security in their organization after participating in cybersecurity initiatives and courses. In this sense, some frameworks provide the dimensions of security culture (see Chap. 4) and the relative metrics in order to evaluate employees' sentiments towards security in their company (e.g. Laycock et al. 2019).

Finally, we must consider the measurement of the tangible effects of the training on the company (level 4), such as the increased productivity or the improved quality and efficiency. For this goal, data and information coming from different internal areas and comparing current outcome to previous periods should be used. Moreover, the ROI (return on investment) is one of the indicators used to evaluate the relationship between costs and benefits of the training activity. We know that it is not an easy thing to calculate the ROI for cybersecurity awareness programmes, especially because we are dealing with security awareness.[10] In this case, a standard mathematical equation does not appear to be the most appropriate evaluation method; on the contrary, a combination of approaches could be more effective. For instance, considering what discussed in this paragraph, positive results—such as the reduced number of phishing emails opened a greater attention in reporting phishing attempts—can help to estimate the value of the investment in cybersecurity employees training.

Finally, we underline that when positive outcomes are shared, they can have a motivational impact on the entire organization: making people aware of the results achieved with their contribution is the best measure to ensure their sustained participation to CSC programmes.

## References

Abiddin, N.Z.: Mentoring and coaching: the roles and practices. J. Human Resour. Adult Learn., 107–116 (2006)

Amorim, J.A., Hendrix, M., Andler, S.F., Gustavsson, P.M.: Gamified training for cyber defence: methods and automated tools for situation and threat assessment. In: NATO Modelling and Simulation Group (MSG) Annual Conference 2013 MSG-111 (2013)

Anderson, M.E., Shannon, A.L.: Toward a conceptualization of mentoring. J. Teach. Educ. **39**, 38–42 (1988)

Black, T.: Doing Quantitative Research in the Social Sciences. Sage Publications ltd, London (1999)

Black, P.E., Scarfone, K., Souppaya, M.: Cyber security metrics and measures. In: Wiley Handbook of Science and Technology for Homeland Security, John Wiley and Sons (2009)

Blythe, J.M.: Cyber security in the workplace: understanding and promoting behaviour change. In: Proceedings of CHItaly 2013 Doctoral Consortium, pp. 92–101 (2013)

Cai, Y.: Using case studies to teach cybersecurity courses. J. Cybersecur. Educ. Res. Pract. **2018**, 2, Article 3 (2018)

Carr, L.T.: The strengths and weaknesses of quantitative and qualitative research: What method for nursing? J. Adv. Nurs. **20**(4), 716–721 (1994)

Chew, E., Swanson, M., Stine, K., Bartol, N., Brown, A., Robinson, W.: Performance Measurement Guide for Information Security. National Institute of Standards and Technology, Gaithersburg, MD (2008)

Corradini, I., Nardelli, E.: Building organizational risk culture in cyber security: the role of human factors. In: Ahram, T., Nicholson, D. (eds.) Advances in Human Factors in Cybersecurity, AHFE 2018. Advances in Intelligent Systems and Computing, vol. 782. Springer, Cham (2019)

Council Directive 2008/114/EC, 8 December 2008 on the identification and designation of European critical infrastructures and the assessment of the need to improve their protection, Official Journal of the European Union (2008)

Crookall, D., Saunders, D.: Towards an integration of communication and simulation. In: Crookall, D., Saunders, D. (eds.) Communication and Simulation: From Two Fields to One Theme (pp. 3–29). Clevedon, UK: Multilingual Matters (1989)

ENISA: Cyber Security Culture in organisations (February 2018). https://www.enisa.europa.eu/publications/cyber-security-culture-in-organisations

E-learning Market Report—Global Outlook and Forecast 2019–2014 Report (2019)

[10]https://thedefenceworks.com/blog/is-it-possible-to-calculate-an-roi-for-security-awareness-training/.

EY, Cybersecurity incident simulation exercises (2017)
Gallwey, W.T.: The Inner Game of Tennis. Jonathan Cape, London (1975)
Garris, R., Ahlers, R., Driskell, J.E.: Games, motivation, and learning: a research and practice model. Simul. Gaming **33**(4), 441–467 (2002)
Greitzer, F.L., Kuchar, O.A., Huston, K.: Cognitive science implications for enhancing training effectiveness in a serious gaming context. J. Educ. Resour. Comput. **7**, 1–10 (2007)
Hamari, J., Koivisto, J., Sarsa, H.: Does gamification work?—a literature review of empirical studies on gamification. In: Sprague, R. et al. (ed.) 47th Hawaii Int. Conference on System Sciences, pp. 3025– 3034. Hawaii, USA (2014)
Hamblin, A.C.: Evaluation and Control of Training. McGraw Hill (1974)
Hendrix, M., Al-Sherbaz, A., Victoria, B.: Game based cyber security training: are serious games suitable for cyber security training? Int. J. Serious Games **3**(1), 53–61 (2016)
Ives, Y.: What is 'coaching'? An exploration of conflicting paradigms. Int. J. Evid. Coaching Mentoring **6**(2), 100–113 (2008)
Kaufman, R., Keller, J.M., Watkins, R.: What works and what doesn't: evaluation beyond Kirkpatrick. Perform. Instr. **35**(2), 8–12 (1995)
Kianpour, M., Kowalski, S., Zoto, E., Frantz, C., Øverby, H.: Designing serious games for cyber ranges: a socio-technical approach. In: 2019 IEEE European symposium on security and privacy workshops (EuroS&PW), pp. 85–93 (2019)
Kirkpatrick, D.L.: Training and Development Handbook: A Guide to Human Resource Development. McGraw-Hill, New York (1976)
Kirkpatrick, D. L.: Great ideas revisited: revisiting Kirkpatrick's four-level model'. Train. Dev. **50**(1), pp. 54–57 (1996)
Kirkpatrick, D.L., Kirkpatrick, J.D.: Implementing the four levels. A practical guide for effective evaluation of training programs. San Francisco, CA: Berrett-Koehler (2007)
Knowles, M.: The modern practice of adult education: Andragogy versus pedagogy. Rev. and updated ed. Englewood Cliffs, NJ: Cambridge Adult Education (1980)
Knowles, M.: Associates: Andragogy in action: applying modern principles of adult learning. San Francisco: Jossey-Bass (1984)
Kraiger, K., Ford, J.K., Salas, E.: Application of cognitive, skill-based, and affective theories of learning outcomes to new methods of training evaluation. J. Appl. Psychol. **78**, 311–328 (1993)
Laycock, A., Petric G., Roer K.: The Seven Dimensions of Security Culture, CLTRe (2019)
Le Compte, A., Elizondo, D., Watson, T.: A renewed approach to serious games for cyber security. In 7th International Conference on Cyber Conflict: Architectures in Cyberspace (CyCon), pp. 203–216 (2015)
Maennel, K., Ottis, R., Maennel, O.: Improving and Measuring Learning Effectiveness at Cyber Defense Exercises', Secure IT Systems. pp. 123–138 (2017)
Micallef, N., Arachchilage, N.A.G.: Changing users' security behaviour towards security questions: A game-based learning approach. Proceedings of MilCIS, IEEE (2017)
Minter, R.L., Thomas, E.G.: Employee development through coaching, mentoring and counseling: a multidimensional approach. Rev. Bus. **21**(1/2), 43–47 (2000)
Nagarajan, A., Allbeck, J.M., Sood, A., Janssen, T.L.: Exploring game design for cybersecurity training. In: Proceedings of the 2012 IEEE International Conference on Cyber Technology in Automation, Control, and Intelligent Systems (CYBER): 256–262. May 27–31, 2012, Bangkok, Thailand (2012)
Noesgaard, S.S., Ørngreen, R.: The effectiveness of e-learning: an explorative and integrative review of the definitions, methodologies and factors that promote e-learning effectiveness. Electron. J. e- Learn. **13**(4), 278–290 (2015)
Paivio, A., Rogers, T.B., Smythe, P.C.: Why are pictures easier to recall than words? Psycho. Sci. **11**, 137–138 (1968)
Payne, S.C.: A Guide to Security Metrics. SANS Institute Information Security Reading Room (June 2006)

Punch, K.F.: Introduction to Social Research: Quantitative and Qualitative Approaches, 1st edn. SAGE Publications, London, Thousand Oaks California, New Delhi (1998)

Osula, A.M., Maennel, O. (eds.): Proceedings of the 5th Interdisciplinary Cyber Research Conference 2019, Tallin University of Technology (29th of June, 2019). https://www.taltech.ee/public/k/kyberkriminalistika_ja_kyberjulgeoleku_keskus/proceeding.pdf

Ragins, B.R., Cotton, J.L.: Mentor functions and outcomes: a comparison of men and women in formal and informal mentoring relationships. J. Appl. Psychol. **84**, 529–550 (1999)

Rantos, K., Fysarakis, K., Manifavas, C.: How effective is your security awareness program? An evaluation methodology. Inf. Secur. J.: Global Perspect. **21**, 328–345 (2012)

Sauvé, L., Renaud, L., Kaufman, D., Marquis, J.S.: Distinguishing between games and simulations: a systematic review. Educ. Technol. Soc. **10**(3), 247–256 (2007)

Setola, R., Mastrapasqua, M., Tesei, M., De Maggio, M.C., Corradini, I., Pantaleo, C., Capitello, M.E., De Simio, F.: Study on Security Awareness in Gas Infrastructure. Final report, NITEL (2015)

Tamkin, P., Yarnell, J., Kerrin, M.: Kirkpatrick and Beyond: a review of training evaluation. The Institute for Employment Studies, U.K., Report Number 392 (2002)

Whitmore, J.: Coaching for performance: Growing Human Potential and Purpose—The Principles and Practice of Coaching and Leadership, 4th edn. Nicholas Brealey Publishing, London (2009)

Wilson, M., Hash, J.: Building an Information Technology Security Awareness and Training Program, National Institute of Standards and Technology (NIST), Special Publication 800–50, Computer Security (October 2003)

Xu, K., Yao, D., Pérez-Quinones, M.A., Link, C., Scott Geller, E.: Role-playing game for studying user behaviors in security: a case study on email secrecy. In: Collaborative Computing: Networking, Applications and Worksharing (CollaborateCom), 2014 International Conference On, 18–26, IEEE

Young, M.: Experiential Learning = Hands-On + Minds-On". Market. Educ. Rev. **12**(Spring), 43–51 (2002)

Zyda, M.: From visual simulation to virtual reality to games. Computer **38**(9), 25–32 (2005)

# Conclusions

Over the last few years the theme of cybersecurity has widely grown. Considering the rapid increase of digital technologies, it is most likely that the issue will be even more critical in the years to come.

Data breaches, cyberattacks and security incidents occur more and more frequently; individuals and organizations will have to improve their capabilities to handle them. Indeed, criminals will continue to exploit any possible vulnerability, whether it is technological, physical or human.

We should have the courage to recognize that the current approach to cybersecurity, essentially based on a technocentric point of view, does not work. The evidence of this is publicly available: we only have to read the various international reports on cybersecurity and the numerous data breaches analysis reported in the media.

Despite the unproductive results deriving from the current approach, we observe a persistent trust in the capabilities of digital technologies. They are seen as a miracle able to solve many problems and challenges of our time, to the point where vendors and media continue to stress their advantages, often neglecting the relative risks.

In handling cybersecurity risks, we should not forget the central role played by humans and their relationship with digital technologies. We do not want to criminalize technologies, but it is a fact that we are becoming less and less autonomous in taking decisions and our tendency is to rely more and more on technology.

Differently from the past, we are now dealing with powerful technologies, such as Artificial Intelligence, on which we have even less control. In this sense, lack of human control can represent a new serious problem, affecting security, too. In addition, ethical and social questions have to be faced, since these technologies have a significant impact on our lives.

It is time to change our approach to cybersecurity radically, and to consider the human factor in the right perspective. People are considered the weakest link in the security chain, since most data breaches and incidents involve human element. Instead, humans can become the strongest link for any organization when they are well-trained and motivated to participate to security activities.

© Springer Nature Switzerland AG 2020

I. Corradini, *Building a Cybersecurity Culture in Organizations*,
Studies in Systems, Decision and Control 284,
https://doi.org/10.1007/978-3-030-43999-6

Involving individuals in the building of cybersecurity culture means, above all, understanding their needs, overcoming certain prejudices, and investing in serious educational programmes. If the radical change of paradigm discussed in this book is not adopted, we are convinced that all efforts to effectively handle cybersecurity are bound to fail.

Human beings are the key point, regardless of the technological solutions adopted. Who thinks that the employment of AI solutions will completely replace human beings in the activity of protection has not yet understood the nature of the issue.

Underestimating the importance of the human factor is the biggest threat to an effective cybersecurity.

The question is: is your organization going in the right direction?

The manufacturer's authorised representative in the EU is Springer Nature Customer Service Centre GmbH, Europaplatz 3, 69115 Heidelberg, Germany. If you have any concerns regarding our products, please contact ProductSafety@springernature.com

Printed and bound by CPI Group (UK) Ltd, Croydon, CR0 4YY
22/09/2026
02221057-0018